Chrisanthi

TECHNOLOGICAL CHANGE AT WORK

TECHNOLOGICAL CHANGE AT WORK

Ian McLoughlin and Jon Clark

OPEN UNIVERSITY PRESS
Milton Keynes · Philadelphia

Open University Press
Open University Educational Enterprises Limited
12 Cofferidge Close
Stony Stratford
Milton Keynes MK11 1BY

and

242 Cherry Street
Philadelphia, PA 19106, USA

First published 1988

British Library Cataloguing in Publication Data

McLoughlin, Ian
    Technological change at work.
    1. Employment. Effects of technological change
    I. Title   II. Clark, Jon, 1949–
    331.12′5

    ISBN 0–335–15417–4
    ISBN 0–335–15416–6 Pbk

Library of Congress Cataloging-in-Publication Data

McLoughlin, Ian.
    Technological change at work/Ian McLoughlin and Jon Clark.
       p.   cm.
    Bibliography: p.
    Includes index.
    1. Labor supply — Great Britain — Effect of technological
    innovations on.   I. Clark, Jon, Ph.D.   II. Title.
    HD6331.2.G7M39 1988
    331.12′0941—dc 19   87–36848 CIP

    ISBN 0–335–15417–4
    ISBN 0–335–15416–6 (pbk.)

Typeset by Burns and Smith, Derby
Printed in Great Britain by Redwood Burn Limited, Trowbridge, Wiltshire

Contents

List of figures

List of tables

Acknowledgements

This book was inspired by, and to a large extent draws upon, research and teaching in which we were involved as members of the New Technology Research Group (NTRG) at the University of Southampton. As the research was a collaborative exercise much of the material and many of the ideas in this book are a product of joint efforts with other members of the Group, particularly Howard Rose, Robin King, John Smith, Patrick Dawson, Heather Rolfe and Audrey Jacobs. The Group's work was conducted under a research programme funded by the Joint Committee of the Science and Engineering Research Council (SERC) and Economic and Social Research Council (ESRC), whose support we would like to acknowledge. In addition Ian McLoughlin would like to thank the ESRC for providing the Postdoctoral Fellowship that supported his research on the introduction of Computer Aided Design (CAD) in engineering drawing offices. Naturally we would also like to thank all the managers, union officials and employees in the various organisations involved in the research for their time and efforts which made the studies possible.

In writing the book a number of further acknowledgements are due. We would like to thank Ian Beardwell and Patrick Dawson for their comments on parts of the manuscript, Stephen Barr, formerly Commissioning Editor at the Open University Press, for encouraging the initial idea of the book, and John Skelton of the Open University Press for his part in the final stages of the book's production. In addition we would like to thank colleagues who have supplied us with unpublished and forthcoming material, and the Kingston Business School at Kingston Polytechnic and the Department of Sociology and Social Policy at the University of Southampton for their support during the book's preparation. We would also like to thank: Basil Blackwell, Cambridge University Press, Croom Helm, the Engineering Industry Training Board, Gower, the Institute of Manpower Studies, Longman, the National Computing Centre, Open University Press, Oxford University Press, Frances Pinter, and the Policy Studies Institute, for permission to reproduce illustrations and tables. Above all we would like to thank Kim and Georgiana for their greatly appreciated support throughout.

Finally we would like to acknowledge our debt to all of the members of the NTRG, past and present, and to thank them for their friendship, help and occasional hindrance over the past few years – some of it might have been possible without them but none of it would have been as much fun!

Ian McLoughlin and Jon Clark
Hampton and Southampton
April, 1988

Abbreviations

ACARD	Advisory Council for Applied Research and Development
ACTT	Association of Cinematograph, Television and Allied Technicians
AEU	Amalgamated Engineering Union (formerly AUEW)
APEX	Association of Professional, Executive, Clerical and Computer Staff
AUEW	Amalgamated Union of Engineering Workers (now the AEU)
AUEW–TASS	Amalgamated Union of Engineering Workers (Technical, Administrative and Supervisory Section, now MSF)
ASLEF	Associated Society of Locomotive Engineers and Firemen
ASTMS	Association of Scientific, Technical and Managerial Staffs (now MSF)
ATM	Automated Teller Machine
TASS	(Formerly AUEW–TASS, now MSF)
BIFU	Banking, Insurance and Finance Union
BT	British Telecom
BR	British Rail
CAD	Computer Aided Design
CAM	Computer Aided Manufacture
CAMT	Computer Aided Measurement and Test
CAPP	Computer Aided Production Planning
CIM	Computer Integrated Manufacture
CAD/CAM	Computer Aided Design/Computer Aided Manufacture
CNC	Computer Numerical Control
CSE	Conference of Socialist Economists
ESRC	Economic and Social Research Council
EDP	Electronic Data Processing
EFTPOS	Electronic Funds Transfer at Point of Sale
EETPU	Electrical, Electronic, Telecommunication and Plumbing Union
ENG	Electronic News Gathering

EPOS	Electronic Point of Sale
GLC	Greater London Council
GLEB	Greater London Enterprise Board
GMB	formerly GMBATU
GMBATU	General, Municipal, Boilermakers' and Allied Trade Union
IKBS	Intelligent Knowledge Based System
ISDN	Integrated Services Digital Network
IT	Information Technology
LAN	Local Area Network
FMS	Flexible Manufacturing System
MAP	Microprocessor Applications Project
MDI	Manual Data Input
MDI-CNC	Manual Data Input – Computer Numerical Control
MIS	Management Information System
MSF	Manufacturing, Science and Finance Union
NALGO	National Association of Local Government Officers
NC	Numerical Control
NCC	National Computing Centre
NCU	National Communications Union (formerly the POEU)
NGA	National Graphical Association
NTA	New Technology Agreement
NUPE	National Union of Public Employees
NUR	National Union of Railwaymen
PABX	Private Automatic Branch Exchange
POEU	Post Office Engineering Union (now the NCU)
PSI	Policy Studies Institute
SERC	Science and Engineering Research Council
STE	Society of Telecom Executives
TGWU	Transport and General Workers' Union
TOPS	Total Operations Processing System
TSSA	Transport Salaried Staffs' Association
VDU	Visual Display Unit
WIRS	Workplace Industrial Relations Survey

Introduction

The case of Fortress Wapping[1]

On Saturday 25 January 1986 a group of journalists from the *Sun* newspaper gathered outside the Tower Hotel in London. According to a news report in the *Financial Times*, the journalists had voted the previous evening by 101:8 to accept a package worth £2000 a year, plus membership of a private health scheme. The alternative, their Editor had said, was the sack. 'For Christ's sake,' he had urged, 'take it with both hands.' At 10 a.m. a group set off from outside the hotel on a coach journey. Their destination was a new printing plant a few hundred yards away at Wapping. The plant had been built at enormous expense and amidst intense secrecy and rumour, ostensibly to print a new evening newspaper *The London Post*. The plant's tight security and high barbed wire topped fences – it was known as 'Fortress Wapping' – suggested other motives. Negotiations had been taking place over a four-year period between News International, owned by the Australian-born media magnate Rupert Murdoch, and the printing industry trade unions. The previous day talks had broken down yet again and 5500 print workers – some of whom were alleged to be earning well in excess of £20000 a year – had been called out on strike. Murdoch's response was to dismiss the strikers without compensation.

The coach carrying the journalists at first headed towards East India Dock and continued for more than a mile before the semi-hysterical hilarity of the passengers persuaded the driver, apparently unfamiliar with the geography of east London, to turn back. The previous week part of another of Murdoch's titles, *The Sunday Times*, had been printed at the plant. It incorporated a twelve-page 'Innovation Special' which included an editorial claiming that Wapping contained the latest newspaper technology and represented the 'future of Fleet Street'. As it turned out, the plant's newsroom was equipped with a 'direct-input' system, the first of its kind in British national newspapers, which enabled journalists to type their copy directly into the computer. Under the terms of previous agreements covering computerisation, the powerful craft union the NGA had managed to retain this work for its members. At Wapping the installation of 'direct-input' technology, although suspected, was never confirmed by News International. In fact, the computer system had secretly been assembled and tested in a 'mock up' of the new newsroom at a warehouse in Woolwich and then smuggled into the Wapping plant.

When they arrived at Wapping the journalists were checked in amid heavy security. For many it was their first visit to their new place of work and they were astounded at what they found.

> The entire floor had been laid out as a fully electronic newsroom, with everything in place; a desk was ready for all of them with a [computer] terminal on each; phones were in place; the switchboard had been installed; and all around were genial men and women, many of them American or Australian, who welcomed them to their new workplaces.
>
> (*Financial Times* 27.1.86)

Subsequently the *Sun* journalists were to be joined by colleagues from all of News International's other titles, and, with the help of an alternative labour force, set about producing the papers from their new location. By moving to a new 'green field' site Rupert Murdoch had succeeded in the most dramatic fashion in introducing new technology on his, rather than the printing unions' terms, something other Fleet Street proprietors had failed to do for over a decade. According to his arch-rival Robert Maxwell, the Fleet Street 'gravy train' on which the unions had ridden for decades had finally 'hit the buffers'.

Automate or liquidate ?

'Automate or liquidate' has been the cry heard frequently in British industry in recent years as the pace of technological change has increased and the competitiveness of firms declined. It is widely believed that if organisations are to be commercially successful then, because of the inevitable competitive benefits that it brings, technological progress must not be resisted. This view is linked, paradoxically perhaps when expressed as it frequently is in newspaper editorial columns, to analyses of the 'British disease'. It is often argued that a major symptom of this ailment is the obstructive role of the trade unions, exemplified *par excellence* by those in national newspapers. It is asserted that unions and their members have a 'Luddite' attitude to technological change. This involves, among other things, the self-interested defence of outmoded technology, working practices and skills, resulting in over-staffing and highly inflated wages, all of which ultimately threaten the competitiveness of the British economy. In short, trade unions and their members on the shop floor are a barrier to technological progress and commercial success. In contrast, when seeking to adopt new technology, employers and management are acting in the economic interests of the nation as a whole, as well as those of their own organisations.

As will be seen in this book, there is little hard evidence to support this often articulated but highly simplistic interpretation of the role of employers, management, trade unions and workforce in the process of technological change at work. For example, a recent survey of UK manufacturing industry suggested that lack of people with appropriate skills, the general economic situation, and the high costs and lack of finances for development, are far more important barriers to technological innovation than union or shop floor resistance (see Figure A.1, Appendix). Findings such as this indicate the need for a much more detailed investigation of the actual influence of management, unions and workforce when new technology is introduced.

The 'automate or liquidate' approach also begs some rather important wider questions. For example, can technology and technological progress really be regarded as an autonomous and neutral force independent of the interests of those who decide to initiate it? Are its positive effects – for example, improved competitive performance – inevitable, and its more negative consequences – for example, reductions in employment – unavoidable? Can for instance, News International's adoption of the latest printing technology be explained entirely in terms of commercial and technological pressures, or were other motives also involved?

A set of ready-made answers to these questions is provided by some Marxist writers (see for example, Braverman 1974; Zimbalist 1979; CSE 1980; Albury and Schwartz 1982; Cockburn 1983; Thompson 1983). They see technological change as being determined by the logic of the historical development of capitalist economies which require employers and their management agents to secure profits by centralising control and deskilling work. In national newspapers this has involved automating the jobs of one of the most powerful, well organised, and best paid sections of the workforce – the time-served, craft-skilled compositor. On this analysis, rather than being a rational response to commercial and technical pressures, it is the employers' need to control work and the deployment of labour that is the 'driving force' behind technological change. It follows that technology is not an autonomous and neutral phenomenon, but rather a political or class weapon in the conflict between capital and labour.

The perspective we will adopt in this book is not meant as an all-embracing alternative to either the Marxist or the 'automate or liquidate' approach. It would be foolish to believe that there are not instances, and some, though perhaps not all, parts of the British national newspaper industry may well be candidates, where the pressures of competition and technological progress are such that change is an irresistible option if a business is to survive. Similarly, only the most naïve observer could deny that there are instances when new technology is both designed and used with the explicit intent of wresting control over work from highly skilled groups of employees – again a charge on which management in national newspapers may in some cases be proud to be found guilty. However, the contention of this book is that neither of these interpretations in itself provides an adequate basis for a full explanation and understanding of technological change at work. Not least because, although drawing attention to the imperatives, pressures and constraints acting upon them, neither of these approaches takes into sufficient account the role of managers, trade unions, and workers *within* organisations in shaping outcomes at critical points in the process of change.

The focus and structure of the book

The central argument to be presented in this book starts from the assumption that the process of technological change in an organisation involves a number of distinct stages. At each stage, issues arise which involve actors in making choices that are likely to be constrained by a range of 'external' factors and also contested and modified over time by other organisation members. In other words, technological change involves a process of choice and negotiation which, within certain constraints, offers scope for managers, unions and workforce to play a significant role in determining whether change occurs at all, and if it does, how it is implemented and what its outcomes are.

The specific focus is on technological change in its most pervasive and significant contemporary guise in Britain – new computing and information technologies. We will be concerned only with the introduction of these new technologies into the production process, that is 'process innovations', rather than technological changes in particular products or 'product innovations' (see Willman 1987; 3–6). There has been much written on the 'technological revolution' that this new technology represents. Since the late 1970s an ever-growing literature on the social implications of new technology has also emerged (see for example Forester 1980, 1985, 1987). However, as others have already noted (see for example Gill 1985; Francis 1986; Boddy and Buchanan 1986), much of this literature has tended to be highly general and futuristic in orientation, and its implications for policy and practice often a matter of 'off-the-cuff' prescription. A major difficulty has been that many commentaries have been based on speculative prediction rather than hard evidence derived from studies of the actual experience of introducing new computing and information technologies. As such, little clue is given to the day-to-day realities of technological change at work and how managers, trade unions, workgroups and individual employees might make more effective contributions in implementing and using new technology. One objective we set ourselves, therefore, was to draw together the findings of the latest and most up-to-date academic research that has sought to focus on the *actual experience* of introducing new technology, in the anticipation that an account and evaluation of these findings may have more immediate relevance to students and organisational practitioners.

The book is inspired by, and to a significant degree is based upon, research that we have been involved in over the past few years as members of the New Technology Research Group at the University of Southampton. This research programme was designed specifically to investigate the process of technological change at workplace level, in particular the role of managers, unions and workforce, and the influence of new technology itself, on changes in work tasks and skills, job content and work organisation, and the way in which work is supervised and controlled.

Whilst, as already mentioned, we also draw on a wide range of other studies which have investigated these and similar issues, it is important to recognise possible sources of bias and areas of omission in the text. First, the book is focused on changes *within* organisations. It does not deal with more 'macro' questions such as the implications of new technology for overall levels of employment, or specific questions such as its implications for divisions between male and female employment, or health and safety at work – although some aspects of these issues are touched upon at various points. Secondly, much of the research material is British in origin. At present there is very little detailed comparative data available on how technological change involving new technology has occurred within organisations in different national contexts. Our focus is therefore firmly on the British experience, although reference to changes in other countries is made at appropriate points. Finally, an attempt has been made to use case-study research to provide detailed accounts of the complex nature of the process of change in particular organisations. This inevitably involves some selectivity in choosing the case study examples. However, where possible, attempts have been made to establish the nature of more general trends and patterns by referring to survey data.

The book is structured as follows. Chapter 1 explores in some detail the significance of computing and information technologies. Is this technology really 'new', and if so, in what sense? In addition it assesses the changes in some of the principal sectors of employment in which this technology is being introduced, and the likely pattern and extent of its diffusion throughout the economy. Chapter 2 considers three contrasting sociological perspectives which have been developed to analyse technological change at work. These are 'contingency' theory, 'labour process' theory, and the 'social action' or 'strategic choice' approaches. The chapter concludes by outlining our own approach which draws eclectically on all three perspectives. This highlights the role of managers, unions and workforce in choosing and negotiating the outcomes of change within organisations, but also suggests that the independent influence of technology itself has to be taken into account as a complement to this analysis. Chapter 3 examines the prime importance of the process of managerial decision-making when new technology is introduced. In particular it considers the question of how choices made at various stages are influenced by senior management strategies. Chapter 4 questions the view that trade unions can simply be understood as a 'barrier' to innovation, and explores their influence on the process and outcomes of change, in particular the relationship between the introduction of computing and information technology and collective bargaining.

Chapter 5 outlines our own approach to the analysis of 'technology' as an independent variable, and analyses in detail the influence of new computing and information technologies on the 'core of work' – tasks and skills. This approach is then used to assess the key capabilities and characteristics of computing and information technologies and their independent influence on work tasks and skills. Chapter 6 takes this analysis a step further by showing how the grouping together of tasks and their allocation to jobs is a matter of social choice and negotiation. Chapter 7 completes this discussion by exploring the manner in which choice and negotiation have shaped forms of control and supervision around computing and information technology. Chapter 8 draws together the arguments and findings discussed in previous chapters and suggests possible options for management and unions in developing a more effective response to technological change and a more effective use of new technology.

Notes

1. The following description of events at 'Fortress Wapping' is based on various newspaper reports and in particular, Lloyd, J, and Hague, H. 'Murdoch wins first round in the battle over Wapping', *Financial Times*, 27.1.86, and Linda Melvern's very readable account of the technological revolution in national newspapers that has been inspired by Rupert Murdoch's News International, *The End of the Street*.

CHAPTER 1

New technology at work

Since the end of the eighteenth century the process of industrialisation has been bound up with technological change. In this sense there is nothing 'new' about the idea of 'new technology'. However, in recent years journalistic, government and academic commentators have viewed the latest developments in computing and information technologies as sufficiently distinctive, dramatic, and far-reaching to warrant the title 'the new technology' (see, for example Barron and Curnow 1979; Sleight *et al.* 1979; Rada 1980; Forester 1980; 1985; 1987). Some writers have gone so far as to claim that we are in the midst of a 'technological revolution' which is not only 'confined to the world of science and technology', but is also 'bringing about dramatic changes in the way we live and work – and maybe even think' (Forester 1985: xii). This book is concerned with one aspect of the implications of this 'technological revolution'. That is the introduction of new computing and information technologies into production processes, or 'process innovation', at work.

The objective of this chapter is to explore in more detail the significance of this 'new technology' – is it really 'new', and if so, in what sense? Some of the areas of work where it is being applied are also examined, and predictions about the likely pattern and extent of its diffusion through the British economy assessed. The broad conclusion reached is that the latest developments in computing and information technologies constitute a distinct stage in the development of process automation. However, the predicted pattern of diffusion of the new technology suggests that its most 'revolutionary' implications in the world of work are, at least as far as Britain is concerned, still at an early stage.

The new technology and the automation of work

In order to assess the significance of computing and information technologies it is useful to place them in the context of long-term developments in the automation of production processes. Broadly speaking, since the Industrial Revolution three phases of automation can be identified (see Gill 1985: 63–4; Coombs 1985):

1 *Primary mechanisation*, the use of machinery driven by steam power to replace

human physical and manual labour in the transformation of raw materials into products.

2 *Secondary mechanisation,* the use of machines powered by electricity to accomplish the transfer of materials between machines and to run continuous-flow assembly lines and processes.

3 *Tertiary mechanisation,* the use of electronics-based computing and information technologies to co-ordinate and control transformation and transfer tasks.

Primary mechanisation was predominant up to the end of the nineteenth century, while during the first half of the twentieth century, secondary mechanisation was most significant. Since World War Two, aided by developments in electronics, tertiary mechanisation has assumed increasing importance as more and more aspects of the control of work operations have been automated.

Tertiary automation involves the convergence of three lines of technological development – telecommunications, computing and electronics. In the past, telecommunication systems have been based on the use of 'analogue' signals, but increasingly they are incorporating electronic or 'digital' techniques (Forester 1985: xiii). Since this is the same language as that used by computers there are new possibilities for linking computer systems and developing new telecommunication and information services (see Mayo 1985).

The principal capabilities of computing and information technologies can be described as information capture, storage, manipulation, and distribution (Buchanan and Boddy 1983: 10–11). *Information capture* refers to the gathering, collection, monitoring, detection and measurement of information. This can be accomplished 'actively' through automatic electronic sensors, monitors or process controls, or 'passively' where humans input information. *Information storage* involves the automatic conversion of numerical and textual information into binary digital form for retention in an electronic memory from which information can be retrieved when required. *Information manipulation* involves the automatic organisation and analysis of numerical, textual and graphical material. *Information distribution* involves the automatic transmission and display of information on visual display screens or paper, and the exchange of information between machines or computer systems.

It is these capabilities that give computing and information technologies the capacity to automate the control of work operations. This can happen in at least three ways (Buchanan and Boddy 1983: 12):

1 The equipment can give to the operator feedback information to make operator control of the equipment or process more effective.

2 The equipment or process can be taken under computer control through a predetermined sequence or cycle of operations.

3 Deviations from equipment or process standards can be measured and corrective action initiated by the computer.

As well as aiding the control of work operations, new information and computing technologies can also promote organisational integration in a number of ways: by the greater accessibility of information to different levels and areas; the speed with which information can be communicated between organisation members; and new possibilities for the display of performance information at central points. Similarly, increases in management control may be enabled by faster and more precise knowledge on work operations; reductions in the scope for 'indeterminate behaviour' by subordinates (e.g. 'figure adjusting' in reporting performance information); and opportunities for previously separate control systems to be unified allowing managers to make more comprehensive and balanced assessments of work performance (Child 1984: 257–8).

It is the 'control' capabilities of computing and information technologies which identify them as a distinct phase in the automation of work, and for this reason 'tertiary automation' is sometimes referred to as 'control automation'. However, it is important to realise that even where the latest technology is incorporated into these systems, the control of work operations is not completely automated in the literal sense that no human intervention is required at all. As Buchanan and Boddy note, it is more realistic to adopt Bright's definition of automation as meaning 'more automatic than previously existed', when considering apparently new stages in technological change (Bright 1958; quoted by Buchanan and Boddy 1983: 12). As will be seen in subsequent chapters, the realisation that automation does not mean the total replacement of human labour in the control of work operations is essential if the implications of new computing and information technologies are to be properly understood.

The significance of microelectronics

The principal feature of developments in tertiary automation in the post-war period has been the growing application of increasingly powerful, low-cost, and smaller electronic components (see Noyce 1980). The invention of the microprocessor or 'silicon chip' in 1971 marked the most significant stage yet in this process. Microelectronics radically transforms the capabilities of computing and information technologies and extends the range of their applications in both products and in the production process itself. It also makes it possible to innovate in areas of the control of work operations which hitherto could not be 'automated'. Microprocessor-based applications of computing and information technology can therefore be seen as the latest phase in 'tertiary' or 'control automation' (see Coombs 1985).

Microelectronics technology brings a number of benefits to computing and information technologies. It radically enhances their *processing power*; it increases the *speed* with which information is processed and calculations are made; it greatly reduces their *size*; it substantially improves their *reliability*; it increases their *flexibility*; and perhaps most significantly, it reduces greatly the *costs* in relation to processing power. For example, according to the Club of Rome, in the fifteen years up to 1983 the processing power of computers had increased 10000 times, while the cost of each unit of performance had decreased 100000 times (Friedrichs and Schaff 1982). In a more graphic illustration, Forester points out that the first electronic computer – 'ENIAC' – built in the United States in 1946:

> ...weighed 30 tons, filled the space of a two car garage, and contained 18,000 vacuum tubes, which failed on average at the rate of one every seven minutes. It cost half a million dollars at 1946 prices.

In comparison, today:

> ...the same amount of computing power is contained in a pea-sized silicon

chip. Almost any home computer costing as little as $100 can out perform ENIAC.

(1985: xiii)

In order to harness the substantially improved capabilities of computer hardware, a new industry has rapidly emerged devoted to the writing of software programs to exploit these benefits (see *Business Week* 1985; Forester 1987: 146–51).

These improvements in performance coupled with reduced size and cost, and the growing availability of increasingly powerful and sophisticated software, mean that the range of potential applications of microelectronics is enormous, 'unleashing', according to Forester, 'a tidal wave of technological innovation' (1985: xiii). Before considering the nature of the innovations 'unleashed' in the world of work by microelectronics it is necessary to look in a little more detail at a possible explanation of why the technology should be considered to be so significant.

Microelectronics: a new 'heartland' technology?

The potential benefits promised by microelectronics have supported the growing belief that, if the introduction of computing and information technologies based on this technology proceeds quickly enough, then major competitive advantages will be secured for a nation's economy. One argument for taking this view is provided by the theory of 'long waves' of economic development (see Freeman *et al.* 1982; Freeman 1986). Put very simply, the idea of 'long waves' suggests that, while inventions are occurring all of the time, they are transformed into successful innovations only in very specific historical periods. These 'waves' of innovative activity are related to long-term economic cycles of boom and slump which were originally identified in the 1920s by a Soviet economist, Kondratieff. He observed that it took the world economy about fifty years to move through a boom-to-slump-to-boom cycle. Subsequently economists have traced the development of 'Kondratieff cycles' or 'waves' into the latter part of the twentieth century.

Research conducted by the Science Policy Research Unit at Sussex University (see Freeman *et al.* 1982) has shown that economic upturns in Kondratieff cycles are stimulated by the widespread diffusion of 'new technology systems' in a sort of 'bandwagon' effect. Initially, pioneering firms adopt a new technology in the design of their products and new firms and industries 'spin off' from these 'pioneers' to manufacture and sell these new products. They are subsequently bought and adopted as process innovations by 'mature' firms and industries who do not have the expertise to develop the new technologies themselves – hence the 'bandwagon effect' as more and more firms adopt the technology (see 1982: 64–81). The technology underlying such cycles or waves of innovative activity can be regarded as a 'heartland' technology in so far as it supports related families of innovation, firms and industries.

According to long-wave theory, the world economy in the mid-1980s is near the bottom of the slump in the '4th Kondratieff cycle' and experiencing a severe economic recession (see Figure 1.1). Many economists and innovation theorists have therefore devoted their attention to identifying the kind of innovations which may conceivably stimulate the next 'upturn' or '5th Kondratieff cycle' (see Open University 1986a; Hall 1986; Ray 1986 for discussion). Naturally, microelec-

Figure 1.1 Kondratieff Cycles and Innovation Waves

(a) Innovation peaks in trough of Long Wave

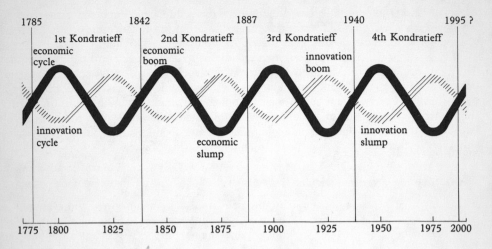

(b) Innovation accelerating during up-swing of Long Wave

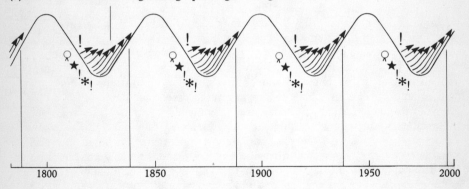

Source: Open University (1986)

tronics, along with other contemporary innovations such as biotechnology, has been viewed as just the kind of 'heartland' technology which could be the basis for starting the 'bandwagon' rolling into the next Kondratieff 'upturn' (see Kaplinsky 1984; Ray 1986; Freeman 1986). Many policy-makers and commentators have thus been encouraged to stimulate and support the development and application of microelectronics, in the hope at least that this will generate the kind of innovation needed to support a renewed phase of sustained economic growth.

Indeed, following in the wake of an upsurge in media interest in Britain during the late 1970s, the adoption of microelectronics-based innovations has been a major preoccupation of successive governments, as well as employers' organisations and trade unions. The overriding concern has been that the new technology is not being adopted rapidly enough relative to our competitors (see for example, ACARD 1979). Accordingly attempts have been made to promote its adoption. In 1978, for example, the then Labour government launched a Microprocessor Applications Project (MAP), subsequently continued by the Conservative government, to stimulate awareness and provide financial support for firms seeking to adopt the new technology in their products or production processes. A survey at the time of the scheme's inception found that only fifty per cent of the firms questioned had even heard of microelectronics (see Braun 1986)!

As will be seen in Chapter 2, one problem of perspectives such as 'long-wave theory' is that it tends to assume that technology has unavoidable 'impacts' on society and work organisations, for example that the decision to adopt a new technology necessarily leads to competitive advantage and in turn stimulates economic growth. However in reality, as much of the empirical research discussed in this book will show, such advantage is determined, not by technical capabilities and characteristics alone, but also by the nature of the process of change within adopting organisations. It is the processes *within* organisations, as well as the 'impacts' upon them, which are decisive in shaping the outcomes that result from the introduction of technology.

To summarise so far, computing and information technologies can be regarded as a distinctive stage in the development of process innovations at work. Microelectronics technology, when incorporated into computing and information systems, radically increases their processing power, speed, reliability and flexibility, while decreasing both cost and size. The range of potential applications of computing and information technologies is therefore increased. Indeed, so radical are the implications of microelectronics that it has been regarded as the kind of 'heartland' technology which could stimulate a renewed phase of sustained economic growth in advanced industrial economies such as Britain. Governments, therefore, have been particularly keen to promote the adoption of this new technology in the belief that competitiveness and future economic growth are dependent upon its widespread use.

'Chips with everything'?: computing and information technologies at work

Where are the potential applications of these 'new' technologies in production processes ? The list appears to be almost endless. According to Ray:

> The microprocessor is a chameleon...it takes the character of whatever program has been fed into it; it can direct a guided missile, operate a coffee dipenser, regulate the use of petrol in a car or control an industrial process. If properly programmed it can be used almost anywhere, in communications, in metal machining, and in widely varying applications, from libraries' bibliographies to medical diagnosis.
>
> (1986: 276)

Not for nothing, then, have headline writers referred to 'chips with everything'! It would be impossible here to review in detail all the potential applications of new technology in production processes. The following pages will therefore concentrate on the main applications of computing and information technologies in three sectors of employment where some of the most 'revolutionary' implications of the technology have been predicted – design and manufacturing, administration and offices, and retailing and finance (see Figure 1.2).

Manufacturing: towards the workerless factory?

Two principal areas of application can be identified in manufacturing employment – design and drawing offices and the shop floor. Computers have been used to aid the design process for several years. Hitherto their principal role has been in numerical analysis to assist in the creation, modification, analysis and optimisation of design choices. However, in the 1960s computer systems began to be developed which utilised interactive computer graphics to display data in the form of pictures and symbols on visual display units (VDUs). It thus became possible to draw using electronic tools rather than the traditional manual tools of pencil, T-square and a

Figure 1.2   Examples of computing and information technology in design and manufacture, offices and administration, and finance and retailing

DESIGN AND MANUFACTURE
* Computer Aided Design and Drafting (CAD)
  - quantitative analysis
  - interactive computer graphics
* Computer Aided Manufacture (CAM)
  - computer numerical control (CNC)
  - robotics
  - flexible manufacturing systems (FMS)
* Computer Integrated Manufacturing (CIM)
  - CAD/CAM
  - computer-aided production planning (CAPP)
  - computer-aided measurement and test (CAMT)

ADMINISTRATION
* Computing
  - on-line processing
  - real-time management information systems (MIS)
  - word processors, desk-top computers
  - intelligent knowledge-based systems (IKBS)
* Telecommunications
  - electronic mail
  - viewdata and on-line data bases
  - private automatic branch exchanges (PABX)
  - local area networks (LANs)

FINANCE AND RETAILING
* Finance
  - automated teller machines (ATM)
  - electronic funds transfer (EFT)
* Retailing
  - electronic point-of-sale machines (EPOS)
* Finance and Retailing
  - electronic funds transfer at point of sale (EFTPOS)
  - integrated circuit cards ('smart cards')

drawing board. The first 'CAD' systems required large and expensive main-frame computers and therefore were developed only for specific applications in defence, aerospace and electronics. However, advances in microelectronics have meant that smaller, cheaper and more powerful minicomputers are now available to support these systems. Improvements in screen technology, programming and memory techniques have enabled the design of 'turnkey' CAD systems which can now be purchased 'off the shelf' and with the appropriate software used for a variety of design and drafting applications ranging from electronic engineering through to civil engineering and architectural applications (see Arnold and Senker 1982; Kaplinsky 1982; 1984: 39–54).

The application of computer technology on the shopfloor has been in evidence since the 1950s (see Kaplinsky 1984: 55–76; Forester 1987: 195–217). The principal development has been the use of what is basically an electro-mechanical technique, known as Numerical Control (NC), to control the actions of metal-cutting machine tools. This involves programming the machine using previously prepared punched paper tapes which contain the information necessary for the machine tool to perform cutting actions on a metal workpiece, such as drilling and milling, in the manner of a 'player-piano'. Prior to the adoption of NC techniques, the control of the cutting action made by these machines was the responsibility of skilled craft workers who interpreted design drawings and operated the controls manually. The technical advantages of NC machines over human operators lie primarily in the improved quality of work, reductions in scrap rates, and time savings.

Developments in microelectronics have resulted in a new generation of Computer Numerical Control (CNC) machines with their own microprocessor-based controls which enable programming at the machine and offer far greater flexibility compared with their NC predecessors. For example, the most sophisticated CNC machine tools are 'machining centres' which can perform not one but several types of cutting operations in a pre-programmed sequence. Accompanying developments in CNC technology have been advances in industrial robotics, and most recently Flexible Manufacturing Systems (FMS) (see Ayers and Miller 1985; Bylinsky and Hills Moore 1985).

Robots were first introduced by the American company Unimation in the 1960s. Most robots consist of mechanical arms which can perform a variety of operations. At their simplest they involve rudimentary 'pick and place' tasks. More sophisticated robots can perform a complex sequence of pre-programmed actions. Again, microelectronics offers enhancement of the capabilities of robots and, along with developments in their sensory abilities which enable them to 'see' and even 'smell', extends the range and complexity of tasks which can be robotised. However, the principal use of robots to date has been in environments within manufacturing processes which are 'hostile' to human beings, for example, where health and safety hazards are involved, or tasks are highly repetitious such as in automobile paint-spraying and welding operations.

FMS represents the next major step in manufacturing automation. Here groups of CNC machine tools and robots are linked together by a central computer, which not only contains all the necessary design and manufacturing information to control machining operations, but is also able to control the movement of parts between machines by automatic transfer and handling equipment. The attraction of FMS lies in the greater flexibility it promises to bring to the control of manufacturing operations and the savings achieved from reduced inventories of parts awaiting machining. Perhaps most significant is the longer-term possibility of linking several FMS 'cells' together to provide a completely automated machining process. Such systems have the potential for eliminating the need for direct labour on the machine shop floor.

The next stage of development is the electronic linking of computer-aided design to computer-aided manufacturing systems (CAD/CAM), co-ordinated and controlled by computer-aided production planning systems – the so called 'workerless factory' or 'factory of the future' (see Forester 1987: 170–94). This would allow production information to be transmitted from the design function to the shop floor in electronic form, thus cutting down the need for large numbers of engineering drawings to be produced by the drawing office. Computer-aided production planning (CAPP) systems would assist management in the co-ordination and control of highly automated production facilities, together with computer-aided measurement and testing (CAMT) systems to allow a more accurate and effective quality control. In short, a fully computer integrated manufacturing (CIM) system. According to the Council for Science and Society:

> Ultimately, we can glimpse a future in which the operation of an engineering business will revolve around a conceptual image held in a computer, or shared between a collection of computers. Building a ship will mean creating an image of a ship in a computer through CAD, as a vast set of numbers defining its shape and the shape of all its parts, and all other factors relating to the design, building and operation of the ship. Out of the computer system will come orders for steel plates, instructions to machines which weld them, estimates of completion date, reasons for delays, and so on. Much of the computer system might be the same, whether the company was making ships, cars or typewriters.
>
> (Council for Science and Society 1981: 29)

Administration: towards the paperless office?

Capital investment per employee in offices has remained considerably lower than that on the shop floor throughout this century (for a discussion see Kaplinsky 1984: 77–107; Giuliano 1985). Since the 1950s, automation and computerisation in offices has generally been of the 'back-room' type, where large main-frame computer systems have been used to provide off-line processing power to support clerical, administrative, sales and professional functions. Current changes in the way administrative and other services are provided are based on the application of microelectronics in office computing and telecommunications systems. Microelectronics enables the widespread introduction of on-line micro- and minicomputer systems, and stand-alone or 'desk-top' personal computers (PCs) into offices and administrative operations where computerisation was previously impossible. Within the office, the processing power of computers provides an electronic means of storing and processing textual, graphical and numerical information, for example, through the use of word processing and financial spread-sheet software (see Forester 1987: 131–46; 195–217).

The electronic transmission of information between offices is facilitated by new telecommunications technologies such as electronic PABX ('private') exchanges and local area networks (LANs) which allow office systems to be linked together ('networked'), and the introduction of electronic public switching and transmission systems, such as that currently being undertaken by British Telecom (for a discussion see Clark *et al.* 1988: 33–49; and for more general developments, Forester 1987: 81–130). This permits access to a wide range of other electronic services such as electronic mail, on-line data bases and interactive view data and information retrieval services. Some commentators go as far as to predict that paper could eventually be eliminated as the medium for recording, communicating and storing information in the office (see for example, Cawkell 1980). However, a more realistic experience seems to be that electronic office systems result in the consumption of as much, if not more, paper than their non-automated counterparts. In the view of a manager interviewed by one of the present authors, the 'paperless loo' was more likely to be achieved than the 'paperless office'!

The first computer information systems used in office and administrative work were off-line systems for the routine bulk processing of data in batches. The information that managers received provided a retrospective view of the performance and status of work operations. The next development was in on-line information systems which provided managers with real-time data on performance and operational conditions. Both these types of system were based on large main-frame installations. Typically they were administered by central data-processing departments. This meant that computers were quite often regarded as an interesting but not essential tool by line managers, who were too busy to digest the volume of information they produced since it was presented in long print-outs, difficult to read and understand. Microelectronics has enabled computers to be used as a far more flexible means of accessing and retrieving management information from computer data-bases (see *Business Week* 1985).

Personal computers now provide managers with powerful desk-top systems which can be used for a variety of diagnostic and decision-support tasks. For example, the product of one computer company acts as multi-purpose desk-top microcomputer. This incorporates, among other things, an integral digital telephone, an answering machine which employs a voice synthesiser rather than tape recorder, and various applications software packages. In the longer term, the development of intelligent knowledge-based systems (IKBS) has been predicted (see for discussion Feigenbaum and McCorduck 1985; Weizenbaum 1985; Forester 1987: 45–9). It is anticipated that these systems will be capable of providing support for the qualitative aspects of managerial and professional decision-making, ranging from medical diagnosis to engineering design. Thus computers may increasingly be used in an 'expert' consultant rather than 'number cruncher' role.

One potentially dramatic extension of the use of microelectronics in managerial and professional work is to relocate the 'electronic office' workstations in the home of the employee and link them via the telecommunications system to their former workplaces. This offers to the employer considerable savings on office overheads and an opportunity to redefine the relationship between them and their managerial and professional employees who can be employed on a sub-contract rather than permanent basis. Some commentators see such developments as the forerunner of 'telecommuting' where managers and professionals will work from home rather than travel into the central business districts of large cities. This suggests, among other things, a potentially radical reduction in the size of work organisations. This phenomenon has been likened by one futurologist to the cottage industries of the early industrial revolution, describing the office workplace of tomorrow as the 'electronic cottage' (Toffler 1980). Whether a realistic prediction or not, it is the case that major computer companies, such as ICL and Rank Xerox, have conducted small-scale experiments in 'homeworking' of this type (see for discussion Huws 1984).

Finance and retailing: towards the cashless society?

Microelectronics also offers potential change in the way services are provided in the 'front offices' of organisations – in particular where employees deal directly with the customer. This is most noticeably the case to date in finance and retailing, where 'back office' application of computers has been in evidence since the 1960s (see Marti and Zeilinger 1985; Forester 1987: 218–40). For example, banks and more recently building societies[1], are changing the way they provide services to their customers at branch level. The intention is to reduce the need for direct contact between employees and customers by automating routine transactions, so allowing bank staff to concentrate on more profitable non-routine transactions such as loans and insurance (*Banking World* April 1987). To this end banks and building societies have been installing computer-based systems in the 'front office' lobby area and its environs of their branches, either for the use of tellers when conducting transactions with customers or for direct use by the customers themselves. Of particular note to date has been the widespread introduction of Automated Teller Machines (ATMs, or 'cash dispensers') which allow 'out-of-hours' and 'through-the-wall' banking. The insertion into the machine of a plastic card and personal identification number (PIN) by the customer allows him/her to conduct transactions, such as withdrawals or account balance checks, twenty-four hours a day without the help of a member of staff and often without entering the bank itself (*Euromonitor* 1985; *Banking World* November 1986).

Current developments include more sophistcated ATM machines which can provide a broader range of services to customers (in the USA weather reports and baseball scores can be obtained from some ATMs!) and the 'networking' of ATM machines from different banks and building societies so that customers can use the machines of a number of institutions (*Banking World* March 1986). In addition, experiments are taking place in some banks with 'interactive video disc' technology which allows customers to use their plastic cards and PIN numbers to 'dial-up' on a sophisticated ATM machine, equipped with a computer processor and video screen, both words and pictures to advise them on particular services or products – such as loans and morgages – being offered by the bank. One of the most recent developments has been the concept of 'home and office banking', which allows customers linked to the bank through the public telecommunications system, to carry out financial transactions twenty-four hours a day without even leaving home (*Banking World* November 1986).

Similar trends are evident in retailing with the introduction of 'intelligent' cash register machines known as Electronic Point of Sale (EPOS) terminals. These machines are being used to replace mechanical and electro-mechanical cash registers or 'tills'. Some of the most sophisticated versions are being introduced in supermarkets and incorporate a laser scanner which can read the computer bar-code information found on the packaging of most products. This permits the computation of the customer's bill, and, since the terminals can also be linked to stock-control computers, allows the maintenance of a real-time record of stock levels which prompts the re-order of the depleted product lines (see *Euromonitor* 1985). Additionally, accurate and up-to-the-minute sales information is provided which can be accessed by managers. Again, in the United States things have been taken a step further and some EPOS terminals are equipped with voice synthesisers which wish customers 'a nice day' on completion of their transaction!

A further development is likely to be the electronic linking of the retail system to the banking system. Instead of exchanging cash for goods, it is now possible for intelligent cash terminals to be linked to the banking computer system and for the transfer of funds to be made electronically, Electronic Funds Transfer at Point of Sale (EFTPOS) systems, as they are known, can require only a plastic cash or credit card from customers to pay for their goods. In France a 'smart-card' with an inbuilt microprocessor has been developed which will allow the user to carry out a variety of transactions with only one piece of plastic (see *Financial Times* 20.2.85; *Banking World* December 1986). In future it may be possible that the customer will not even need to leave home, since the whole transaction could be accomplished via electronic links provided by the public telecommunications system (*Euromonitor* 1985; *Banking World* January 1986). Such developments have prompted some commentators to see EFTPOS technology as a step towards the 'cashless society.

New technology at work: a new technological revolution?

How far are the 'technological revolutions' in the world of work implicit in the ideas of the 'workerless factory', 'paperless office' and 'cashless society' actually taking place in reality? This section examines a possible scenario for the pattern of innovation and diffusion of microelectronics through the economies of advanced industrial nations. The subsequent section then looks at available survey evidence on the pattern of innovation and current diffusion of new technology in Britain. This indicates that although significant changes are occurring, the 'technological revolution', if it is to happen, is still at an early stage.

Possible scenarios for the diffusion of microelectronics

It was suggested above that the one reason why microelectronics-based innovations are considered so important is that they may constitute the kind of heartland technology which will support the upswing phase of an economic long wave in a 'bandwagon' effect. Freeman *et al.* (1982) have attempted to plot the diffusion of microelectronics applications, in effect the route of the 'bandwagon', through the economy (see Figure 1.3). As adequate statistics on the diffusion of microelectronics through the economy do not exist, Freeman *et al.*'s model is 'impressionistic'. Nevertheless, the model suggests that the initial and most rapid adoption of the new technology since the early 1960s has been in 'pioneer' science-based industries such as defence, space and some areas of consumer electronics. Since the mid-sixties these pioneering efforts have spread to other 'high-tech' industries such as motor vehicles and machine tools.

Figure 1.3 Diffusion of microelectronics applications

Rate of Diffusion*	Rapid (from 1960)	Medium (from 1965)		Slow (from 1965)		
(Depth of impact)†	High	High	Medium	High	Medium	Low
Design and redesign of products to use micro-electronic technology	Electronic capital goods Military and space equipment Some electronic consumer goods	Machine tools Vehicles Electronic consumer goods Instruments Some toys	Other consumer durables Engines and motors Other machinery	Some biomedical products	Other toys	Agriculture Hotels and restaurants Construction Personal services
Process automation using microelectronic technology	Some electronic products	Machining (batch and mass), especially in vehicles, consumer durables and machinery Printing and publishing	Continous flow processes already partly automated e.g.: • chemicals • metals • petroleum • gas • electricity	Clothing Textiles Food Assembly	Building materials Furniture Mining and quarries	Agriculture Hotels and restaurants Construction Personal services
Information systems and data processing	Specific government, business and professional systems involving heavy data storage and processing in large organisations	Financial services Communication systems Office systems and equipment without total electronic systems Design	Transport Wholesale distribution Public administration Large retailers	Retail distribution All-electronic office systems Electronic funds transfer	Domestic households Professional services	Agriculture Hotels and restaurants Construction Personal services

* Ranging from less than 10 years (rapid) to more than 30 years (slow) for the greater part of production to be affected.

† Proportion of total product or process equipment cost.

Source: Freeman et al. (1982)

The use of microelectronics in the actual design of new technology products such as computers, word processors, robots and machine tools was a necessary prerequisite to their adoption by the mass of other potential users in mature industries who were without the knowledge or skills to develop the technologies in their own production processes. The 'bandwagon' effect in the automation of shopfloor and office work is therefore likely to have begun in the mid-1960s, and started in other areas during the 1970s. Significantly, Freeman *et al.* suggest that in the past the diffusion of major new technology systems to cover even half of the potential adopters has taken more than thirty years. If this model is correct it would suggest that the microelectronics 'bandwagon' may only be in the early stages of its journey, for although applications may have been diffused widely in the products and processes of science-based industries, they are only now being adopted on a large-scale by the mass of potential users in more mature industries.

A similar conclusion is reached by Kaplinsky (1984). He traces the likely diffusion of microelectronics applications in terms of what he sees as three distinct types of control automation. To illustrate this point he suggests that in most modern industrial enterprises there are three spheres of production – design, manufacture, and co-ordination (management and administration) – each with its own sets of activities. Kaplinsky's three types of control automation are distinguished by the extent to which microelectronics applications facilitate the *convergence* of activities into single machines or systems and the *integration* of previously discrete spheres of production. The types of automation identified are as follows (see Figure 1.4):

1 *Intra-activity* automation, where single activities within a sphere of production are automated, for example, in the design sphere when a manual drawing board is replaced by a CAD terminal. This is often termed 'island automation'.
2 *Intra-sphere* automation, where separate activities within a particular sphere of production converge or are linked into a single system. An example of this type of automation in the manufacturing sphere is provided by FMS systems, which bring together the machining, transfer and control of the manufacturing process into a single computer-controlled system.
3 *Inter-sphere* automation, where separate activities or groups of activities in different spheres of production are integrated. An example of this would be a CAD/CAM system. A more sophisticated example would be a fully integrated computer-controlled manufacturing system – in other words a fully automated 'factory of the future' (1984: 24–6).

According to Kaplinsky, intra-activity automation is now widespread in all three spheres of production. Current applications of the new technology are now focused on intra-sphere automation (1984: 65–86). The ultimate 'prize' however is inter-sphere automation. According to Kaplinsky there are technological, social and economic impediments to the achievement of this goal which make predicting a timescale for these developments difficult. However, he suggests: 'the 1980s are likely to see increasing signs of individual sets of inter-sphere automation technologies and by the 1990s we will surely see the fairly widespread emergence of fully automated production in many sectors...' (1984: 106). How far though are these scenarios borne out by the current rate and extent of the diffusion of new technology?

Figure 1.4   The three different types of automation

(a) Intra-activity

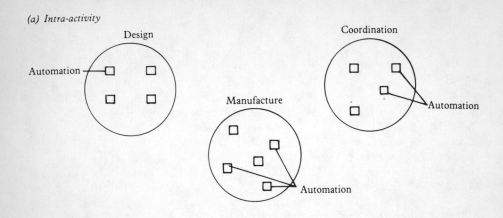

(b) Intra-sphere

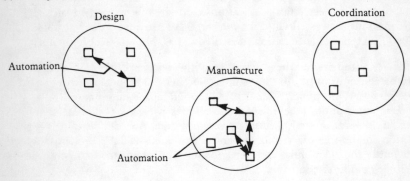

(c) Inter-sphere

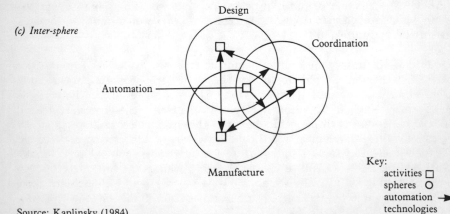

Key:
activities ☐
spheres ○
automation ➤
technologies

Source: Kaplinsky (1984)

The pattern of innovation and diffusion of new technology

The most comprehensive and representative data on microelectronics applications in manufacturing in the UK is provided by three successive surveys conducted by the Policy Studies Institute in 1981, 1983 and 1985 (see Northcott *et al.* 1982; Northcott and Rogers 1984; Northcott 1986). These surveys show that the use of microelectronics in new technology products (product innovations) has been concentrated in science-based industries, in particular electrical engineering, mechanical engineering and vehicles, ships, and aircraft. Use in production processes (process innovations) has, as Freeman *et al.* predicted, spread more widely to all sectors of manufacturing to include firms in more mature industries who do not have the skills or knowledge to develop applications of the technology themselves (see Figure 1.5; also Willman, 1986: 168–74).

The PSI surveys also reveal an increasing rate of adoption of microelectronics in production processes during the 1980s. Whilst in 1981 eighteen per cent of UK establishments with more than twenty employees were using microelectronics in their production process, this had risen to forty-nine per cent by 1985. In 1985, the establishments using microelectronics in production accounted for over three-quarters of total manufacturing employment. However, while this suggests a very rapid diffusion of the technology, the latest survey indicated that the rate of growth in the number of users was beginning to slow. Moreover, many users of the technology do not do so extensively. On average, UK users utilise microelectronics in the control of just under one-third of their production processes. The overall rate of use suggests that about one-quarter of all production processes in the UK are controlled in some way with the aid of microelectronics.

The general pattern of adoption of individual technologies is shown in Figure 1.6. In 1985, CAD work stations were being used by seven per cent of UK establishments, CNC machine tools by fourteen per cent, PLCs (programmable logic controllers) by twenty-seven per cent, pick-and-place machines by three per cent and robots by two per cent. However, when the type of application is considered, the survey revealed that the use of microelectronics in the control of individual machines or processes (forty-one per cent of UK establishments covering sixty-eight per cent of manufacturing employment) is far more common than more advanced applications such as the centralised control of individual machines or integrated control of several stages of processes (four per cent of UK establishments covering six per cent of manufacturing employment). This would suggest that 'system' applications such as integrated CAD/CAM systems and FMS have so far been extremely limited and that 'island automation' has tended to predominate. This is a pattern evident in the previous surveys and is also supported by other research (see for example, Arnold and Senker 1982; Bessant and Haywood 1985; Willman 1986).

A further set of survey material which is of relevance here is from the Workplace Industrial Relations Survey (WIRS) also conducted by the Policy Studies Institute, which included among other things major sections of questioning relating to technological change (see Daniel 1987)[2]. This survey was conducted in 1984 and covered a representative sample of 2000 establishments with twenty-five or more employees drawn from manufacturing and service industries in the public and private sectors, but excluding the coal-mining industry. While not exploring the diffusion of particular technologies in any detail it does, as will be seen in subsequent chapters, provide the most comprehensive and up-to-date survey material available on the relationship between technology and changes in workplace industrial relations[3].

Figure 1.5   Use of microelectronics in UK by industry

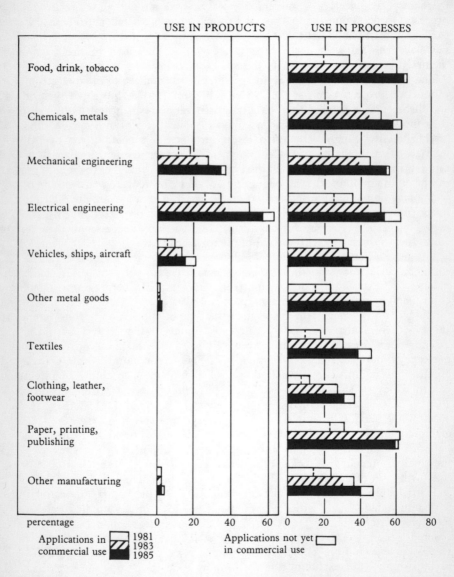

Percentage of all UK factories

Source: Northcott (1986)

Figure 1.6 Types of process equipment used in UK manufacturing

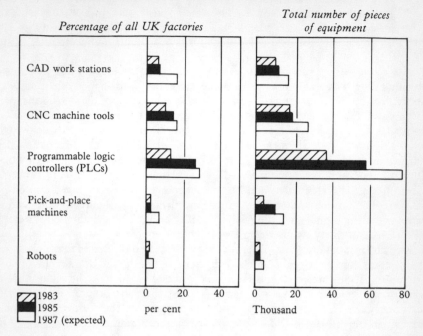

Source: Northcott (1986)

The survey found that microelectronics-based technology has been introduced in fifteen per cent of all workplaces, involving one-third of manual workers. However, in the case of manual workplaces in manufacturing, microelectronics technology has been introduced in over thirty per cent of cases. Because these have tended to be larger workplaces this change has affected over half of manual employees in manufacturing. In fact, the survey reveals that forty-three per cent of manual work-places in manufacturing now use some form of microelectronics application in their production processes (Daniel 1987: 30). The patterns of usage in manufacturing varies according to three factors: size of manual workforce, ownership and industrial sector. Nearly ninety per cent of establishments employing over five hundred manual workers have adopted the technology, and, on average, in each case about one-quarter of the workforce employed are working in the areas of production covered. Second, the use of new technology is much more common in foreign-owned manufacturing establishments than their domestic counterparts, and in establishments which are part of a group rather than independent outfits. Finally, new technology is most common in the engineering and process industries and least common in textiles, clothing and footwear (Daniel 1987: 31–3).

A similar picture is evident in the office, where developments also appear to have been evolutionary rather than revolutionary. A survey in 1985 by the National Computing Centre (NCC) of the uses of electronic office technology in 139 UK firms confirmed a pattern evident in other studies (see for example, Steffens 1983;

Swann 1986). These findings revealed that individual pieces of office equipment such as word processors are far more likely to have been introduced than system technologies such as PABXs and LANs which switch and transmit information between office locations (see Figure 1.7). The NCC survey also suggested that even large firms who are established computer users have been slow to innovate in this way.

However, the survey did reveal a growing sophistication in the variety of uses of electronic office hardware. This reflects the increasing importance of the software industry, which is extending the range and sophistication of the applications programs available for computer systems. The growing pace of innovation in offices, in particular for mini- and micro- and personal computer systems, is indicated by the following figures. Since 1978 the number of main-frame computers being installed has grown at a rate of twelve per cent per annum. However, the rate of growth of small business computers has increased by twenty-seven per cent, word processor systems at seventy-six per cent and business microcomputers by one hundred per cent. Moreover, while the number of current applications is small, communications systems providing digital switching and transmission links have increased at a rate of one hundred and thirty-two per cent a year.

Nevertheless, it has also been observed that the future diffusion of many electronic office systems and services will be dependent not only on the spread of local telecommunications networks and switching systems, but also on the modernisation of the public telecommunications and switching infrastructure. Indeed many observers regard the slow progress made in Britain in this respect as a major constraint on the diffusion of many computing and information technologies. At the time of writing, for example, only 1000 fully electronic telephone exchanges have been installed in Britain, and over 3500 electro-mechanical exchanges have still to be replaced (Clark *et al.* 1988: 47). Many of the 'system' benefits to be gained from a fully electronic telecommunications system (or Integrated Services Digital Network, ISDN, as it is known) can be achieved only once the whole network is fully electronic. In Britain it is not anticipated that such benefits will start to be realised until the early 1990s (see for a discussion, Clark *et al.* 1988: 191–5).

The WIRS confirms the increasing importance of microelectronics applications in the office. The survey found that non-manual workers are in fact twice as likely to have experienced the introduction of new technology than manual workers. Over one-third of offices have experienced technological change in the form of computers or word processors. These workplaces covered nearly two-thirds of non-manual employees. Nearly two-thirds of office workplaces use one or more computing facilities and one-quarter use word-processing equipment.

The extent of use of computer technology again varies according to a number of factors. First, the survey suggests that the larger the size of the non-manual workforce the more likely the use of a variety of computing facilities. Second, the use of computer facilities is very much less common in the public sector than in the private sector. Third, in the private sector, as in manufacturing processes, foreign-owned establishments are more likely to use computing facilities. However, in the case of office employment, independent workplaces are more likely to use the technology than those which were part of a group. Finally, the leading industrial sectors using computing technology are to be found in financial and business services, followed by retail and wholesale distribution. The least likely users are in

Figure 1.7 Penetration of office equipment in 139 UK companies

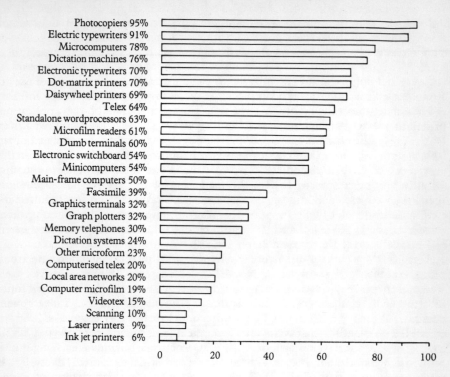

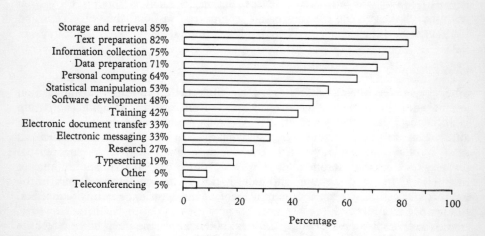

Source: NCC (1986)

hotel and catering (Daniel 1987: 48–52). A broadly similar pattern is evident in the case of word-processing equipment (ibid: 53–5)

Other evidence from the financial and retail sector shows that the major changes required to support the transition to a 'cashless society', although in hand, are still very much at the experimental stage. By the end of June 1987, for example, over 9975 ATM machines had been installed by the major banks, and 1709 by building societies and non-'high street' banks. This compared to only a few hundred far less sophisticated machines a decade earlier, while 1986 saw two major 'networks' of ATMs under the brand names 'Matrix' and 'Link' being established by two groups of building societies (*Banking World* 1986; 1987; see also Marti and Zeilinger 1985; Rajan 1984; Willman and Cowan 1984). However, as already noted, only one bank (the Bank of Scotland) has so far introduced a nationwide 'home and office banking' service which enables customers to transfer funds electronically between accounts without visiting a branch. In fact the marketing of this service has so far been focused on professionals and small business owners rather than private consumers.

The diffusion of EPOS equipment appears to have been rather uneven, being most enthusiastically embraced by large supermarket chains, and to a lesser extent by 'variety' stores. Department stores have shown less interest in the technology. According to one survey, superstore and hypermarket retailers intend to have eighty per cent of their cash-taking equipment EPOS-based before 1990. The survey itself, undertaken in 1984, identifies forty-eight retailers who were using EPOS and found 275 terminals equipped with laser scanners (*Euromonitor* 1985). Latest reports suggest that the penetration of EPOS equipment in retailing by the end of 1986 was around ten per cent, mainly in DIY, electrical and department stores. Planned moves towards the development of an EFTPOS network linking major retail outlets with the banking and finance system have been widely reported in the press. However, it appears unlikely that such a scheme will have any significant impact until the mid-1990s, and even then cash is still expected to remain the predominant medium of exchange. In Britain the first major steps towards EFTPOS will be taken with an experiment involving the installation in retailers' premises of 2000 terminals, linked to the clearing banks' computers, in Southampton, Leeds and Edinburgh by 1989 (*Financial Times* 5.3.88).

Conclusion

This chapter has attempted to assess the claims that computing and information technologies, in particular where they involve microelectronic applications, are 'revolutionary' and 'new' and that they can be seen as a distinct stage in the automation of work. It has been suggested that microelectronics can be regarded as a new 'heartland' technology which marks a new phase in the development of 'tertiary' or 'control' automation. The wide range of potential applications of microelectronics in computing and information technologies was illustrated by looking at uses in three areas of employment: manufacturing, administration, and finance and retailing.

However, despite the seemingly limitless potential for applying the new technology, the extent of its current diffusion indicates that the new 'technological revolution' predicted by some commentators, if it is to occur, is as yet at an early stage in Britain. This returns the discussion to some of the questions raised in the introduction to the book. In particular, will these new technologies have inevitable and unavoidable implications for work and organisation, or will the changes that occur be determined by other factors too? Chapter 2 sets out to explore a number of sociological perspectives which have sought to deal with these questions, and outlines our own approach to the analysis of contemporary technological changes at work.

Notes

1 An immediate consequence of the Financial Services Act 1986, which took effect on 1 January 1987, has been to enable Building Societies to provide a number of other services in direct competition with banks. The building societies have only half the number of branches and so are seeking to create electronic networks to deliver their new range of services to the market place (*Banking World* November 1986).
2 The survey was first conducted in 1980 and is intended to be repeated every four years. In 1984 the survey was sponsored by the Department of Employment, the Economic and Social Research Council, the Policy Studies Institute, and the Advisory, Conciliation and Arbitration Service.
3 One interesting feature of the survey is that it allows a comparison of the application of microelectronics-based innovations with other forms of change affecting work between 1981 and 1984, namely, organisational change not involving technology, and conventional technological change not involving microelectronics. The survey suggests that the most common form of innovation that has affected manual workers in the early 1980s has been organisational change. This is followed closely by conventional technological change, followed closely by advanced technological change involving microelectronics. Office workers have not only experienced more change involving microelectronics than their manual counterparts, but in general have also tended to experience more change of the other types as well. Overall, just under half of office workplaces, accounting for over seventy per cent of non-manual employees, had experienced change involving new equipment or organisation during this period (Daniel 1987: 14–21; 39–45).

CHAPTER 2

Analysing technological change at work

In the introduction to this book, reference was made to the widely held view that technological progress poses a simple choice for management, unions and workforce. They either 'automate or liquidate'. This popular view finds some support in the academic school of thought known as 'innovation theory', of which the long wave theory discussed in Chapter 1 is an example. In general, innovation theory argues that the introduction of new technology is driven by commercial and technological imperatives (see Wilkinson 1983 for a discussion). Where the 'impact' of a new technology is found to provide an economic advantage to an adopting firm, then other firms are forced by commercial pressures to adopt the technology in order to be able to compete and survive.

The rate of diffusion of a new technology is determined by a number of factors. For example, the attitudes of management and unions in firms who are potential adopters may be such that change is resisted – the case for many years, some would argue, in national newspapers. One reason for resistance might be the perceived negative 'impacts' of the technology such as job losses. These, however, are usually treated as adjustment problems which management in particular has to resolve, for example, by educating the workforce in the commercial realities confronting the organisation. In general the tendency, according to Wilkinson, is to view technology 'as a neutral input to individual production systems, the motivation behind its introduction being purely competitive, and its effects, apart from the improvement of the competitive position of the firm or nation-state, being largely incidental' (1983: 9).

The idea that technology has inevitable and unavoidable 'impacts' on organisations was also implicit in many social scientific studies published in the 1950s and 1960s. Much of this work arose largely in response to what had been termed the 'first automation debate' (Benson and Lloyd 1983: 73) which accompanied the initial introduction of electronic computer technology into the workplace in the 1950s. Attempts were made to show how the technology determined such things as work organisation, levels of grievance behaviour amongst work groups, the form of management control, and the degree of alienation or involvement of the workforce within the enterprise (for discussion see Rose 1978: 175–223; Silverman 1970:

100–18; Hill 1981: 85–123). As Rose notes, what unites the wide variety of writers who have pursued such ideas is the desire to 'endow' technology with 'a primary explanatory value' (1978: 212).

These approaches tend to see the introduction of technology as a 'fact of industrial life' which poses problems of organisational adjustment and adaptation for management, unions and workforce, and which require the development of appropriate policy responses (see for example Somers *et al.* 1963; Hunter *et al.* 1970). More recently similar themes have emerged once more in much of the literature which has sought to generalise and predict the social 'impact' of new computing and information technology (for example, Martin 1978; Bell 1980; Stonier 1983; and for a discussion see Lyon 1986).

The aim of this chapter is to examine three influential perspectives on technological change which differ in two main respects from most of those mentioned so far. First, their focus is on technological change *within* the organisation, and in particular at the workplace, that is, they attempt to explain micro- rather than macro- processes and events. Second, their interest and approach is mainly sociological in orientation. They are concerned to varying degrees with the social factors determining organisational behaviour, and the processes through which organisational actors attempt, within certain constraints, to shape organisation structure, the nature of the work and the employment relationship. In the concluding sections of the chapter we will outline our own approach to technological change at work, drawing eclectically on the three perspectives we have discussed.

The chapter begins with a consideration of the work of Joan Woodward writing from the perspective offered by 'contingency' theory. Woodward's is a classic, though sometimes misunderstood, example of an approach which has sought to analyse technology as an independent explanatory variable. In her view technology is a contingent factor exercising a primary influence on organisation structure and behaviour. Two perspectives critical of the 'contingency' approach which seek to establish alternative models and stress the social shaping of technology and change in organisations are then considered. These are labour process theory, which draws on Marxist perspectives and seeks to explain technological change in terms of capitalist imperatives to control labour; and the 'strategic choice' approach, which draws on the 'social action' perspective and is based around the idea that organisations are an emergent outcome of the actions of their members. In fact, these perspectives offer contrasting views on the extent to which the outcomes of technological change are chosen and negotiated by organisational actors or determined by wider commercial, technical and historical forces. They also provide contradictory views of the effects of the new computing and information technology at work.

Contingency theory: technology as a determinant of work and organisation structure

Joan Woodward's work provides a highly influential but much criticised approach to analysing technological change which views technology as an independent explanatory variable, and in some observers' opinion also attributes to it a primary explanatory status. Woodward's ideas were based on over ten years of empirical research during the period of the 'first automation debate' in the 1950s and 1960s (see Woodward 1970, 1980). The starting point for her argument was the theoretical framework provided by 'contingency theory'. 'Contingency theory' advances the proposition that there is no 'one best way' to organise and manage production. Rather, different approaches are appropriate to particular situations depending upon a range of 'contingent' factors such as product markets, labour markets, organisation size, and technology.

Woodward's contribution to this approach was to focus on 'technology' as the most significant contingency shaping organisation structure and behaviour. Using this approach, Woodward's analysis led her to the conclusion that given technologies require management to adopt particular forms of organisation if their enterprises are to be commercially successful. In this analysis she developed a model of the relationship between advanced types of automation and organisation structure which is of considerable contemporary interest. She suggested that advanced production technologies are used most effectively where a specific form of management control system is adopted, which in turn has implications for the content of work and relations between management and employees. In short, she argued that advanced technological change would lead to 'mechanical' and integrated forms of management control which were incorporated within the technology itself. This would relieve management of the need to direct and personally supervise the workforce, whose performance would be subject to control by machinery. How then did Woodward reach this conclusion?

Technology, management organisation and commercial success

In order to identify different types of 'technical situation' it was first necessary for Woodward to define what was meant by 'technology'. She argued that 'technology' could be defined as the 'production system' used by an organisation. This comprised two main elements: equipment, and the assumptions, goals and policies underlying the production process which decide, for example, whether an organisation makes standardised mass products or 'one-off' bespoke products. From her empirical studies Woodward identified eleven different types of production system, which were grouped into three main categories for analysis; unit and small batch production, large batch and mass production, and automated continuous-process production (see Woodward 1980: 39).

To explore the organisational implications of different types of technology it was also necessary to define more precisely what was meant by 'organisation'. This was defined as the management control system employed by the enterprise. When a firm decides to make a particular product or series of products, Woodward argued, a control system is automatically brought into existence. This provides a framework for determining objectives, the sequence of activities required to achieve these objectives, the actual execution of these plans, the assessment of results, and the taking of any necessary corrective action (1980: 189–90).

Woodward's empirical studies allowed her to construct a typology of different types of control system based on the degree to which management control was exercised on a personal or impersonal (mechanical) basis, and the degree to which control systems were integrated with each other or fragmented across the organisation (see Woodward 1980; Dawson and Wedderburn 1980: xix–xxii). Four ideal types of control system were identified (see Figure 2.1). At the one end of the scale

Figure 2.1 Types of management control system

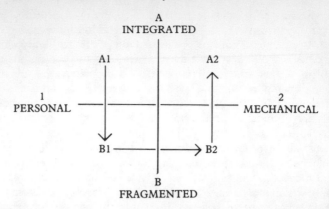

Source: Reeves and Woodward (1970)

were integrated personal control systems (A1), often manifested in the person of an individual owner-manager. In the middle were fragmented control systems (B1 and B2), where standards were set by separate independent departments within the organisation. At the other end of the scale were integrated impersonal control systems (A2), where behaviour was controlled by administrative procedures or by technical devices ('mechanical controls') built into the production system.

Woodward's studies revealed a strong association between types of production system, management control system, and commercial success (see Table 2.1). Firms which had a unit or small batch production system were most likely to be commercially successful if employing an integrated personal control system, while those with large batch or mass production systems were most successful if they employed fragmented control systems of a personal or impersonal variety. Finally, commercially successful firms employing advanced process production systems were most likely to be those who had developed management control systems of the integrated impersonal variety (Woodward 1970: 37–56; 1980: 68–72; Dawson and Wedderburn 1980: xiii). Woodward's overall conclusion, therefore, was that 'there is a particular form of organisation most appropriate to each technical situation' (Woodward 1980: 72).

Table 2.1 The relationship between production systems and management control systems

	Control system A1 %	Control system B1 %	Control system B2 %	Control system A2 %
Unit and small batch production	75	25	–	–
Large batch and mass production	15	35	40	10
Process production	–	–	5	95
Total firms	28	21	18	33

Source: Reeves and Woodward (1970)

Advanced automation, work and management control systems

It was Woodward's belief that technological change would lead organisations to develop control systems in the direction of the arrows in Figure 2.1 – that is, towards the integrated/impersonal control systems 'appropriate' to highly automated production systems – typified at the time she wrote by the continuous process technology used in oil and chemical refineries. To the extent that new technologies give mass and batch production systems characteristics similar to continuous-process production (Child 1984), Woodward's observations on the relationship between highly automated production and management control systems are of considerable contemporary interest. Her contention is that automated production systems demand that impersonal/integrated management control systems be brought into being if the technology is to be used effectively (i.e. to result in 'commercial success'). An important effect of these control systems is to give rise to forms of work which increase the autonomy of work teams, transform the traditional role of the supervisor, and enable a reduction in the potential for conflict between management and labour, thereby offering managers the opportunity to pursue new approaches to industrial relations (see Gallie 1978: 13–14).

With automated production systems, she argues, the key operations involved in transforming raw materials are incorporated within the technology itself. The work tasks of human operators, therefore, are to monitor the operation of the process, keeping the machinery or plant running at as near to full capacity as possible. Unlike in less-advanced types of production systems in which work is broken down and divided into its constituent elements, this work involves operators working in small teams to monitor the production process. In addition, integrated control systems involve establishing a clear and precise balance between competing managerial objectives at the design stage, that is prior to the installation and operation of the plant. In other words, the framework for determining the objectives, the sequence of activities required to achieve these objectives, the actual execution of these plans, the assessment of results and the taking of any necessary corrective action have to be agreed between managerial functions in order that the mechanisms involved can be built into the production system itself.

This again is in direct contrast to the fragmented control systems associated with less-advanced production systems. Here control is exercised independently by various departments, which often results in workers being subjected to contradictory pressures from different management functions, for example, managerial demands to improve both the quantity and quality of production simultaneously. With automated production systems, the required system of management control means that these pressures are absent and the work team has considerably more autonomy to regulate its efforts within the parameters set by the design of the production system itself.

With automated production systems, too, the actual mechanisms for the management control of work operations are impersonal, that is they are incorporated within the technology itself. These 'mechanical controls' mean that there is no longer a requirement for supervisory management to engage in 'policing' the workforce through direct personal control of their work. Once more this is in contrast to less-advanced production systems which require direct supervision and subdivision of work operations to maintain standards. Woodward argues that the absence of direct supervision means that it is therefore possible for managers to devote more time and resources to developing closer and more human relations with the workforce. This means that the role of the supervisor itself is changed by the introduction of advanced technology, requiring supervisors to focus less on the traditional and often conflict-ridden tasks of controlling labour, and more on providing a source of technical expertise, support and coordination to the work groups under their control.

In sum, Woodward's research suggests that the impact of highly automated production systems will be: a reduction in tasks involving direct human intervention in the transformation of raw materials; their replacement with new tasks requiring workers to monitor the work process; forms of job content and work organisation based around work teams; and the erosion of the need for personal supervision which is instead replaced by 'mechanical controls' incorporated within the machinery itself. Overall, suggests Woodward, these changes will lead to a situation 'conducive to the development of harmonious and contributive social relationships' (Woodward 1980: 199, quoted by Gallie 1978: 13). She explains these changes in terms of the technological and commercial imperatives associated with the adoption of advanced production systems. In order to achieve commercial success, managers have to adapt organisation structure to provide the required type of management control system. According to Woodward, therefore, of all the contingent factors which determine organisation structure and behaviour it is the independent influence of the production system (technology) employed which is most significant.

Criticisms of the contingency approach

Whilst, as will be seen below, Woodward's views on the impact of advanced production processes are open to considerable debate, the principal concern of most critics is the method by which she reaches these conclusions. The hallmark of Woodward's approach is the apparent belief that technology can be regarded as the primary independent variable determining forms of organisation. As indicated in the introduction to this chapter, this idea was also implicit in the work of many other writers including innovation theorists and those who have attempted to predict in general terms the 'social impact' of new computing and information technologies. What follows, therefore, applies both to Woodward's ideas, and what have more generally been called 'technological implications' or 'impacts of innovation' approaches.

In Woodward's case, one set of criticisms has come from other exponents of the contingency theory approach who have questioned the primacy she gives to technology and suggest instead that other contingent variables are of more significance (see for discussion Dawson and Wedderburn, 1980 also Dawson 1986). However, the most forceful set of criticisms of both Woodward and the general idea that technology acts as a primary independent explanatory variable are to be found outside the 'contingency' approach. These critics argue that the whole idea that 'technology' has independent 'implications' or 'impacts' on organisations is untenable (the following draws mainly on Wilkinson 1983: 11–12; see also Rose 1977; Silverman 1970; Mackenzie and Wajcman 1985). First, this leads to 'technologically determinist' types of explanation which ignore the social context and processes behind the introduction and operation of technology. Second, insufficient weight is given to the importance of managerial choice in decisions over the introduction and use of technology. Third, the approach assumes a model of the employment relationship which is characterised by a consensus of interest between management and workforce thus 'depoliticising' the issues raised by technological change.

In relation to the first point, critics have argued that in the real world technology does not appear out of thin air, but is the product of social choices and political processes which shape the form and direction of technological innovation. The point is well made by Wilkinson who, as will be seen below, stresses the political nature of technological change and the importance of managerial choice in decisions over the introduction of new technology:

> The impacts of innovation approach is severely limited to the extent that it treats technology as devoid of its own social and political context. Technology is held to arise somehow from scientific and technical research, and if it so happens to confer competitive advantages then firms must adapt and accept the consequences – the 'impacts' or go bust. The *context* in which technical change occurs is treated as of importance only to the extent that it constrains the changes, or is changed itself by the new technology... The fact that managers may introduce technology with the *intention of transforming* the nature of work simply will not fit the 'impacts of innovation' framework,
>
> (Wilkinson 1983: 11, original emphasis)

The second point of criticism follows from this. Playing down or ignoring the significance of managerial choice in decisions over the introduction of technology tends to cast managers in the role of 'messengers' for the technical and commercial system 'which demands they act in accordance with its inherent logic' (Wilkinson 1983: 19). In other words, managerial choices, and the values and ideas which guide their behaviour, are not seen as creative and open, but are judged solely in terms of whether they act (or do not act) in accordance with given technological and commercial imperatives.

The third point of criticism takes issue with the view that the introduction and operation of new technology does not involve potentially strong conflicts of interest between management and workforce. Woodward, for example, gives little consideration to such issues. In so far as she discusses conflict this is in terms of tensions between different divisions or departments within organisations (for example between production and maintenance) rather than between management and workforce. In general, it is assumed that what is good for the commercial success of the company is good for *all* of its employees. Workforce and trade union resistance to change is therefore seen as a reflection of management's failure to communicate the logic of its plans effectively, rather than as an expression of an opposed interest.

The difficulty with this is not so much that there are no common interests involved in the introduction of new technology but that, as Wilkinson observes, the extent of consensus may be limited or conditional. By ignoring the potential for conflict the issues raised by technological change are 'depoliticised' and technology itself is portrayed as a politically neutral force. The idea that technology is 'a phenomenon controlled by particular people with particular interests and in particular positions of power' is therefore sidestepped. The questions which must be addressed, argues Wilkinson, are: 'Why does the change take on this particular form?' 'which groups of people instigate the changes?', 'who are the prime beneficiaries?' (1983: 12). The remainder of this chapter involves a consideration of perspectives which have sought to address these issues.

Labour process theory : capitalism, technology and management control

In contrast to such work as Woodward's, labour process theory seeks to uncover the social and economic interests which lie behind technological change. Not surprisingly this leads to a radically different set of conclusions regarding the effects of new technology. Labour process theory has its origins in the work of Karl Marx, but its recent re-emergence can be attributed to the publication of a book written by Harry Braverman – *Labor and Monopoly Capital* – published in 1974. Braverman set out to challenge what he saw as the technological determinist assumptions of writers such as Woodward, offering a model of the evolution of modern enterprises which identified the capitalist search for increased profitability and management concern for increased control over work as the driving forces behind technological change. The conclusions reached contradicted the view that automation would inevitably raise the levels of skill and autonomy of the workforce. Rather, Braverman argued, automated technologies were introduced by management with the intention of deskilling the content of jobs in order to increase management control over the labour process. How then did Braverman reach this conclusion ?

Management control and the labour process

Underlying Braverman's argument is the view that there is a fundamental and structurally determined conflict underlying the relationship between capital and labour. Relations between management and workforce, and the question of technological change, need to be seen in the light of this basic fact. For Braverman (1974: 56–63) the employment relationship begins when the employer buys the services of the employee in the labour market. In entering into a contract the employer and employee have only agreed the terms and conditions of the employment relationship. The details of what work they do, when they do it, and how it is to be done have to be resolved day-by-day in the workplace. Braverman argues that in order to get the maximum return on their 'investment' in human labour employers have to maximise their control over the behaviour of employees. To the employer this presents itself as 'the problem of *management*' (original emphasis). While Braverman does not ignore the importance of management in reducing uncertainties in the external commercial environment by pursuing increased profits through better sales, marketing and so on, he claims that 'the essential function of management in industrial capitalism' is 'control over the labour process' (1974: 63).

Braverman's view of the problem of management control is therefore in direct contrast to that of Woodward. She saw the problem of management in terms of how to develop control systems most appropriate to the type of technology within the organisation. She assumed that employers, managers and workforce would have a common interest in establishing the appropriate control systems in order to achieve efficient production and commercial success. Braverman believed that management control systems were one aspect of the wider class conflict between capital and labour and that the introduction of automated technologies and control systems was part of a more general attempt by management to wrest control from the workforce over the labour process.

Scientific management, deskilling and technology

How then, has management sought to gain control over the labour process? The answer to this question lay, for Braverman, in the origins of industrial capitalism. At this time workers (by which he meant largely craft workers) were initially in control of the labour process by virtue of the skills they brought to their work. However, this control represented a direct challenge to the employer's drive for efficiency and profit. According to Braverman the most important response to this challenge was the development at the turn of the twentieth century of the theory and practice of 'scientific management' based on the ideas of Frederick Winslow Taylor. 'Taylorism' advocated the pursuit of industrial efficiency by the complete separation of conception (the mental labour of planning and decision-making) from execution (the exercise of manual labour) in the accomplishment of work tasks. The objectives of this 'Taylorist' approach were twofold. First, job content was to be deskilled progressively as work was broken down into fragmented tasks requiring little mental ability and only physical effort to execute them. Second, the functions of planning and decision-making required to direct and control the execution of work were to be increasingly concentrated in the expanding ranks of management and managerial functions.

Initially, the enforcement of the rigid division of labour advocated by Taylorism relied on organisational and disciplinary mechanisms. However, as the twentieth century developed, automation offered to management a more effective means of control over the labour process which would reduce their reliance on the direct personal supervision of labour normally required under Taylorist methods. Thus, for Braverman, whilst technological change enables improvements in productive efficiency it also facilitates 'the progressive elimination of the control functions of the worker, in so far as possible, and their transfer to a device which is controlled, again in so far as possible, by management from outside the direct process' (1974: 212).

It is important to emphasise that Braverman did not see the use of automation to deskill jobs as an expression of the inevitable impact of technology itself. It is entirely a product of the need to control the labour process in order to increase profits. Under a different social system, he argued, advanced technology would open up the possibility of different forms of job design and work organisation which would benefit the workforce. This would involve workgroups possessing the engineering knowledge required to operate and maintain the technology, and a rotation of tasks to make sure everyone had opportunities to work on both highly complex and routine jobs (1974: 230). In other words, rather than being deskilled, they would retain autonomy and control over the labour process, and advanced technology would be used as a complement to, rather than substitute for, human skills and abilities. For Braverman, such a form of work organisation and control was not conducive to the interests of management in a capitalist society, and therefore could not be brought about without a major political, economic and social transformation.

In sum, Braverman's views are based on a model of the implications of technological change which derives directly from management attempts to seek control over the labour process. In this analysis technology has no 'impacts' beyond those intended by management in accordance with the logic of the historical development of capitalism.

Criticisms: labour process theory after Braverman

Braverman's arguments have stimulated an extensive debate on the nature of work in capitalist societies. They have been particularly significant as a counter argument to the view that skills will be increased and workers will become more autonomous as new computing and information technologies are introduced. However, as with Woodward, the analytical method used to reach these conclusions has also caused great controversy. In very broad terms two kinds of criticism of Braverman's analysis have been presented. First, there are those who are sympathetic to his general approach but who see his version of labour process theory as flawed and in need of further refinement (see for example, Burawoy 1979; Zimbalist 1979; Edwards 1979; Friedman 1977; Storey 1983; Thompson 1983). Second, there are those who have taken a more agnostic or critical stance, and although recognising the value of some of Braverman's insights, do not accept it as an adequate or complete explanatory framework (see for example, Littler 1982; Wood 1982; Littler and Salaman 1984; Knights, Willmott and Collinson 1985; Knights and Willmott 1986a; Watson 1986).

There is no space here to consider all of the issues that have been raised by these criticisms – indeed so many critiques and revisions and further criticisms have emerged that one commentator has suggested Braverman has died 'the death of a thousand qualifications' (Eldridge 1983: 419)! However, it is important to note those which are of direct concern to the analysis of technological change at work.

The first major criticism relates to Braverman's concept of management and management strategy. This rests on the assumption that the imperatives of capital accumulation (that is, the drive for profit) require management to wrest control over the labour process from the workforce, and that one particular strategy, 'Taylorism', is the most appropriate way of achieving this. This argument has been questioned on the grounds that there are other strategies available to management in their quest to control labour, and that these may not involve a deskilling of jobs and reduction in worker autonomy. Friedman (1977), for example, has argued that a Taylorist strategy of 'direct control' may be appropriate for semi- or unskilled 'peripheral' workers, but that a strategy of 'responsible autonomy' which delegates a certain amount of discretion to workers in order to gain their commitment to management goals is more appropriate for skilled 'core' workers who are strategically important to the production process. Friedman insists that management strategies essentially involve choices between one or other approach depending on prevailing product and labour market conditions.

However, other writers sympathetic to Braverman's original position have been disinclined to surrender the primacy he attached to the idea of deskilling. Paul Thompson for example insists that 'deskilling remains the major *tendential* presence within the development of the capitalist labour process' (original emphasis) (1983: 118). However, he argues for a more sophisticated model which recognises the sources of variation and constraint on management attempts to deskill labour. For example, he points to the highly varied form that applications of the new technology are taking, and suggests that the state of product and labour markets, and the strength of trade union organisation, will all limit the extent to which technology can be used by management to deskill work (1983: 111–17). Moreover, he rejects Friedman's argument that 'responsible autonomy' represents an alternative strategy to deskilling. Rather, attempts to 'enrich' jobs in this way represent a control strategy which presupposes that these jobs are already deskilled (ibid. :117).

Other writers have sought to question not only the primacy of deskilling as *the* management strategy, but have also challenged the idea that management control of the labour process, by whatever means, is in fact the motive force behind management concerns. This leads to a fundamental criticism of the concept of 'management' and 'management strategy' underlying labour process theory. For example, Batstone *et al*. (1984, 3–6) argue that, while correctly focusing attention on the relationship between overall business strategies and policies for the management of labour, the labour process approach tends to see the latter as flowing unproblematically from the former. Management (in similar fashion to its role in contingency theory – see above) is seen as kind of 'transmission belt' converting the need to improve profits into strategies to control and exploit labour. This ignores the functional division of labour within management between departments concerned explicitly with labour control (i.e personnel and industrial relations) and those corporate functions (for example, marketing, finance, research and development) which are not. Characteristically, labour management issues do not figure prominently in the formulation of business strategies, reflecting in part the fact that personnel and industrial relations management is often a distinct activity marginal to mainstream corporate planning (see Chapter 3). The idea of management subscribing in unison to one overall strategy which has the sole purpose of securing control over the labour process is, therefore, at best a gross over–simplification of what occurs in practice (see also Wood and Kelly 1982).

The second main criticism of Braverman's approach is his tendency to see the implementation of management strategies as unproblematic. This arises because of his view of management as a homogeneous grouping, and the lack of significance attached by his analysis to the individual and collective response of labour to management strategies. Braverman presents a picture which suggests that the workforce passively accepts the deskilling of its jobs and offers no resistance to, or at least makes little attempt to influence, management plans. Again, this view seems to bear little relation to what happens in reality, and much recent empirical work within the labour process approach has sought to document the many instances of labour resistance and its capacity to modify management strategies through engaging in a 'contest for control' over the labour process (see for example, Zimbalist 1979; Friedman 1977; Edwards 1979).

The third main criticism of Braverman's analysis is his insistence on a model of the employment relationship which views relationships between management and workforce on all issues as conflictual. This leads to the conclusion that the workforce always needs to be 'controlled' by management and ignores both the possibility of common interests, and what has been called 'the organisation of consent' (Burawoy 1979: 27). According to Burawoy different groups of workers, rather than being engaged in a continual conflict with management, actively create the conditions for consent by informal networks and practices at work, as well as assimilating consensual views of their role from the wider society. In the context of introducing new technology, then, it is unrealistic to assume that the only response of labour will be to resist management intentions. Indeed, in many circumstances labour may endorse and actively support management plans.

A further criticism of Braverman's version of labour process theory is his inadequate analysis of gender and the position of women (for discussion see Thompson 1983: 190–6, also Beechey 1982; Knights and Willmott 1986b). Over recent years feminist writers have argued that patriarchal relations, i.e. the overall domination of men over women in all aspects of social life, exercise an independent influence on the 'contest for control' at the workplace. Put another way, it is argued that conflict at work occurs not only between labour and capital, but also reflects inequalities and antagonisms in the wider sexual division of labour between men and women. On this analysis, technology needs to be viewed, not only as a means by which management seeks to control labour, but also as a means by which the subordinate position of women may be maintained by men (see for example, Huws 1982; Barker and Downing 1985; Cockburn 1983; 1985a).

One way in which this may be achieved, it is argued, is in the social definition of what are 'skilled' and 'unskilled' jobs, in so far as work traditionally performed by women is frequently labelled as 'low skilled' and accordingly paid at lower rates. Thus, for example, a management introducing new technology which replaces traditional 'male skills' might seek to accomplish deskilling by recruiting an alternative cheaper female labour force. At the same time, the threatened male workforce may well evoke traditional images of what is 'man's' work in an attempt to retain their employment (see for example, Cockburn 1983; 1985a; Crompton and Jones 1984).

The criticisms highlighted suggest that the effects of technological change may be more complex and ambiguous than Braverman and many other labour process writers have suggested. Braverman and his followers are no doubt correct in their criticism of technological determinism, especially where it leads to conclusions that the introduction of automated technology will inevitably raise the general level of skill and autonomy of the workforce. However, the argument that the effects of technological change can be explained largely if not wholly in terms of the management strategies to deskill work is equally flawed. Even if allowance is made for the modification of management strategy during its implementation by the response of labour, the model still assumes that labour control strategies flow unproblematically from overall business objectives during technological change. In other words, for most labour process writers, the deskilling effects of new technology are as inevitable as the 'upskilling' effects associated with the views of writers such as Woodward. It appears that, stimulating and controversial though it has been, labour process theory, and in particular Braverman's rather mechanistic deskilling thesis, has failed to provide an adequate alternative framework for analysing technological change at work. The next section deals with a third perspective which has sought to come to terms with this problem.

The strategic choice approach: choice, negotiation, and the process of technological change

According to Wilkinson, theories of organisation structure and change have traditionally been divided between 'structural or functionalist' approaches which attempt to explain the characteristics of organisations by reference to such things as technology, size and complexity, and 'action approaches' which 'treat organisational structures as emergent entities dependent upon the conscious political decisions of actors and groups of actors' (1983: 9).

The idea that the form of work and organisations is in fact the product of the 'social action' of organisation members emerged as a predominant theme in organisational theory and industrial sociology during the 1960s (see Rose 1978: 227–67; Silverman 1970: 126–73). More recently the 'social action approach', as it is generally known, has been widely adopted as the basis for analysing technological change at work. The key to this new approach is the assumption that the outcomes of technological change, rather than being determined by the logic of capitalist development, or external technical and commercial imperatives, are in fact socially chosen and negotiated *within* organisations by organisational actors. It therefore follows that rather than there being any uniform tendency when new technology is introduced, the changes that occur are likely to reveal considerable variation between organisations, even where the technology concerned and organisational circumstances involved are the same. This directs attention to processes of change *within* particular organisations and the manner in which managers, unions and workforce are able to intervene to influence outcomes.

The concept of strategic choice

An important early contribution to this social action-based approach was made by John Child (1972) who used the concept of 'strategic choice' as a means of emphasising the role of managerial choice, rather than technology, in shaping work and organisation. To be fair, Woodward in her later work did recognise that management choice had a role in determining the nature of control systems, and that technology could be seen as a means by which such decisions were implemented rather than determining the changes itself (see Hill 1981: 105–6; Dawson and Wedderburn 1980: xxii). Nevertheless, as Dawson notes, while 'she did not adhere to a strictly deterministic view... there is little doubt she was more excited by the discovery of the link [between technology and organisation] than she was in discussing the limitations of her analysis or why particular technologies were chosen' (1986: 81). Child's objective, however, was to go much further and highlight the role of choice in explaining variations in organisation and behaviour at work, and to play down the significance of technology as an independent variable.

Thus, whereas Woodward and other 'contingency' theorists explained variations in outcomes in terms of 'situational' factors, such as type of technology, market situation or an organisation's size, Child argued that these were no more than points of reference which decision-makers took into account when designing organisation structures, setting performance criteria and choosing the markets in which to operate. Child argued that it was 'strategic choices' on the part of decision-makers which were the critical variable in the theory of organisation (Child 1972: 15). More recently Child has attempted to take this idea further and marry his ideas to some of the insights offered by labour process theory (see Child 1985).

The utility of the concept of 'strategic choice' is that it draws attention to the question of who makes decisions in organisations and why they are made. This is a useful antidote to the perspectives discussed above which tend to portray technological change, and forms of work and organisation, as though they were independent of the goals and objectives of organisational actors in particular cases. Child argues, in contrast, that decision-making is a political process whereby strategic choices on issues such as long-term organisational objectives, the allocation of resources and organisational design, are normally initiated and taken by a 'power-holding group' or 'dominant coalition' within the organisation. 'In short,' argues Child,

> ...when incorporating strategic choice in a theory of organisations, one is recognising the operation of an essentially political process in which constraints and opportunities are functions of the power exercised by decision-makers in the light of ideological values.

> (1972: 16)

The idea of strategic choice involves two further elements. First, dominant coalitions do not necessarily comprise the formally designated holders of power within an organisation. Indeed, there could be more than one such coalition, leading to conflict and competition between different management groups. Second, although strategic choices are likely to be made by dominant coalitions of senior managers, they can be modified by other organisational actors, particularly through collective action by the workforce, and by middle managers responsible for implementing senior management decisions (Child 1972: 13–14).

The point that the strategic choices made by senior managers may subsequently be modified by other organisational actors has been taken up by other writers employing this approach (see also Child 1985). According to Wilkinson (1983: 18–19), for example, managers act as 'creative mediators between potential and actual technology'. Their decisions reflect positions of power and influence within an organisation and their particular assumptions and values. It cannot be assumed, he suggests, as Braverman and many labour process writers do, that middle and lower managers will always act in the 'interests of a "capitalist class"' in implementing the strategic choices made by senior managers (1983: 16). Similarly, Buchanan contends that, 'individual managers have vested interests or "stakes" in applications and outcomes of technological change in their organisations. These stakes differ between management levels and functions and are not necessarily consistent with the overall performance goals of the organisation' (1986: 79).

In addition to modification and influence by lower-level managers, strategic choices may also be challenged collectively or individually by workers in so far as trade unions, workgroups and individuals seek to contest and influence managerial choices. In this way, the outcomes of technological change within particular organisations are dependent not only on the mediation of lower levels of managers, but also 'on the way workers respond, adapt and try to influence the outcome' (Wilkinson 1983: 21). Finally, social choice and negotiation over technological change must be seen in the context of organisational arrangements and working practices which have already been decided upon and contested, and are thereby part of the custom and practice of the workplace. Following Strauss *et al.* (1973), Wilkinson refers to this as a ' "negotiated" order of unwritten rules' between workers, managers, supervisors and other organisational actors which influences 'both the type of technology that managers seek to introduce and the social consequences of that technology' (1985: 449).

The process of technological change

This social action-based approach has had considerable influence on recent research – including our own – concerning the introduction of new computing and information technologies (see for example, Buchanan and Boddy 1983; Bessant 1983; Wilkinson 1983; Walton 1986; Sorge *et al*. 1983; Child *et al*. 1984; McLoughlin *et al*. 1985; Davies 1986; Clark *et al*. 1988; McLoughlin 1986; 1987; 1988). Many researchers who have adopted this approach view it as an essential corrective to what is seen as the 'technological determinism' of writers such as Woodward and the inadequacies of the labour process approach. According to Wilkinson for example:

> ... the technical and social organisation of work can best be seen as an *outcome* which has been chosen and negotiated...In this view the design and choice of technology may be seen as the result of socially-derived decisions, and the way in which technology is used can be explained in terms of the political processes – formal and informal – of negotiation, persuasion, bargaining, and so on.
>
> (1983: 20)

It follows, as Wilkinson notes, that technology has no uniform impact on work or industrial relations, since this depends on choice and negotiation during the introduction of new technology within individual organisations (1983: 21). However, as will be seen below, we would argue that whilst the outcomes of change may be socially chosen and negotiated, this does not mean that 'technology' itself does not have an independent influence in enabling and constraining social choices.

A concern with the way the outcomes of change are socially chosen and negotiated *during* the introduction of new technology focuses attention on the *processual* nature of change. As Boddy and Buchanan observe, the 'big-bang' image of technical change in which everything falls into place overnight is not consistent with what really happens in practice. They argue that:

> ... technical change must be seen as a process, not as an event. The introduction of a new machine usually involves a long period of planning and preparation before its arrival; modifying it and improving it once installed; and training staff to use and maintain it. The effective introduction of even comparatively minor technical changes is thus protracted and fragmented, and involves a lot of formal and informal negotiation.
>
> (1986: 9)

According to Wilkinson the process of introducing new technology can be broken down into a number of stages, including the choice, implementation, and debugging of technology. During these stages 'junctures' occur at which managers, trade unions and workers make choices and contest specific outcomes of change, such as skills and work organisation (1983: 21). Bessant calls the scope for choice, or room for manoeuvre in making choices during the process of change, 'design space' (1983: 26).

In an attempt to build on these ideas we have outlined elsewhere a model of the process of technological change that consists of five analytically distinct stages (see Clark *et al.* 1988: 31). These are: initiation, decision to adopt, system selection, implementation, and routine operation. *Initiation* refers to the process by which managers identify and pursue an opportunity for the adoption of new technology. *Decision to adopt* refers to the process leading up to the decision to invest resources in its purchase and introduction. *System selection* denotes the process of in-house design and development of a particular system or equipment, or choice of a system from an external supplier. The fourth stage, *implementation*, embraces the process of introducing the technology in the workplace. This includes both technical and human aspects of installing, commissioning and debugging of the chosen technology and the 'mediating' role of management and union strategies towards implementation. The final stage, *routine operation*, is where the system has been brought into service and a stable pattern of working with the technology has been established.

This model captures the temporal element of technological change. However, in all processes of technological change it can be expected that critical points, or 'junctures' as Wilkinson terms them, will arise at which the choices are made, and the way these are contested by formal and informal negotiation will have a major influence on outcomes. These *critical junctures*, therefore, can be defined as the points at which the temporal *stages of change* intersect with particular *issues* (see below) raised by the introduction of new technology. They offer opportunities where management, trade unions and workers can seek to intervene in order to influence particular outcomes of change (Clark *et al.* 1988: 32).

Any process of change is likely to raise both procedural issues (for example how technology is to be introduced, and who should be involved in its introduction) and substantive issues (for example the nature of skill requirements, the content of jobs, staffing levels, pay and grading, training, etc.). These will require decisions to be made by organisation actors, either by conscious choice and negotiation or by omission or non-decision, at various stages in the process of change. In the context of new computing and information technologies the procedural issues that have come to the fore in recent years have been the *stage* at which unions and workforce should become involved in decision-making over the outcomes of change (see for example, Rush and Williams 1984; Winterton and Winterton 1985), and the different *forms* – provision of information, consultation, traditional bargaining, new technology agreements – such involvement should take (see for example, Moore and Levie 1985; Winterton and Winterton 1985). These issues will be a constant theme throughout the book – and in particular in Chapters 3 and 4. But perhaps of greater interest has been the parallel debate about the substantive issues raised by the introduction of computing and information technologies and the ability of managers, unions and workforce to influence them.

In this context we would distinguish broadly between *employment*, *control* and *strategic* issues. Employment issues are those such as pay, grading, hours and job security which in most western industrial nations are the subject of traditional collective bargaining. Alan Fox has described the relations between management and workforce on these issues as 'market relations' (Fox 1966: 6). In contrast, control issues refer to such things as task and skill requirements, job content, work organisation and supervision (see Wilkinson 1983; Buchanan and Boddy 1983). These issues relate to the work people actually do rather than the terms and conditions under which they are employed. Traditionally, these issues, which Fox described as belonging in the province of 'managerial relations' (1966: 6), have rarely been the subject of formal collective bargaining. Finally, strategic issues such as corporate decisions over the type of products to be made, the location of a business or plant, or the kind of production technology to employ, have tended to be determined by senior management alone, although during the debate on industrial democracy in the 1970s (see TUC 1974; 1979; Bullock Report 1977; Elliot 1978) they were identified as key issues for the achievement of wider union objectives and the extension of workforce participation into management decision-making.

In the chapters that follow we will not be arguing that new computing and information technologies uniquely bring all three of these types of issue to the fore. What we will suggest is that the introduction of these technologies can both *highlight* the significance and inter-relatedness of employment, control and strategic issues, and *challenge* the adequacy of traditional management and union thinking and approaches to handling change. Furthermore, technological change may provide the *catalyst* for the development of new approaches to the conditions under which people are employed, the work that they do, and the manner in which they participate in the overall running of the organisation. However, the relative importance and precise significance of employment, control and strategic issues is likely to vary between individual cases. These variations, therefore, are a matter for empirical investigation.

Technology, choice and negotiation

It follows from adopting this analytical approach that it is difficult to predict with any certainty any uniform effects of technological change in general or new computing and information technologies in particular. Deskilling or upskilling of work are outcomes which result from particular processes of choice and negotiation. They cannot, however, be entirely explained as inevitable and general tendencies generated by capitalist imperatives or as unavoidable 'impacts' of technological change and commercial pressures. The outcomes of change are likely to reveal considerably more variation than is implied in both of these perspectives. If there are any general trends or patterns they are just as likely to be the product of the prevalence of similar values and ideologies among decision-makers, the goals and objectives of organisational actors, and the perceived effectiveness of the strategies and policies that they adopt in particular economic and political circumstances.

One final point needs to be made here. A stress on the importance of choice and negotiation could be taken as suggesting that the capabilities and characteristics of 'technology' have little significance for the outcomes of change. Indeed, in common with the labour process theorists, many writers employing the concept of strategic choice have been at pains to avoid what they see as technological determinist explanations. Buchanan and Boddy, for example, stress that 'the capabilities of technology are *enabling*, rather than determining' and that it is 'decisions or choices concerning how the technology will be used' and not 'the technology' which determine the outcomes of change (1983: 255). Similarly, Wilkinson contends that:

Previous analyses have tended to treat new technology as if it had 'impacts' on work organisation – especially skills – which are inevitable in particular technical and economic circumstances. It is in opposition to this view that technical change is here treated as a matter for social choice and political negotiation, the various interested parties to the change being shown to attempt to incorporate their own interests into the technological and social organisation of work.

(1983: Preface)

However, does the fact that the outcomes of change are socially chosen and negotiated mean that 'technology' has no independent influence?

Whilst labour process and strategic choice approaches may be justified in the view that technology is not the 'primary' explanatory variable determining the content of work and organisation structure, we would question whether it is also right to reject insights such as that implicit in the work of Joan Woodward that 'technology' can have an independent influence. This question is taken up in detail in Chapter 5. Suffice it to say here that a key argument to be developed later in the book will be that an analysis of the independent influence of technology is a necessary *complement* to an examination of the way outcomes of change are socially chosen and negotiated.

Conclusion

This chapter has examined three perspectives on the analysis of technological change at work. First, the work of Joan Woodward was outlined as a classic example of an approach which accorded a primary role to 'technology' as an independent explanatory variable. She argued that commercial success was dependent upon adapting organisation structure (management control systems) to suit the requirements of the technology (production systems). Accordingly, a highly automated technology requires that 'impersonal' management control systems are developed which incorporate control into the technology itself. This results in forms of work which involve the monitoring of, rather than direct intervention in, the production process.

Second, the labour process approach was examined. This stressed that technology did not determine change but was a means by which management sought control over the labour process. Thus, management control systems do not arise from 'thin air', as Woodward appeared to assume, but are the product of the class-based conflict between capital and labour. According to Braverman, the function of management as the agents of capital is to control the labour process. The introduction of new technology is a means to this end. It is used to cheapen labour, deskill job content and remove control over the execution of work from the worker and place it in the hands of management. In a direct contradiction of the predictions of Woodward and others, it is argued that the long-term effect of automation – including the new computing and information technologies – is not to increase the skill and autonomy of the workforce but to deskill and degrade labour.

Finally, the 'strategic choice' approach was examined. This was based around the idea that the outcomes of technological change are the products of social choice and negotiation. This approach suggested that it is the actions of organisation actors – managers, unions and workforce – at critical junctures in the process of change which are critical in shaping outcomes and not exclusively technological, commercial or capitalist imperatives. In other words, the changes which result from the introduction of new technology are profoundly affected by the decisions made by managers and the way these are contested by unions and workforce within individual organisations. The remainder of this book will adopt this perspective as the basis for examining the roles of managers, unions and workforce and for exploring the implications of choice and negotiation for the outcomes of technological change at work.

CHAPTER 3

Management and technological change

This chapter examines empirical evidence on the process of managerial decision-making when new computing and information technologies are introduced. In Chapter 2, contrasting perspectives on the role of management were noted. On the one hand contingency theory and labour process theory tended to see managers as 'messengers' for the wider economic and technical system. In particular, labour process theory tended to see management policies to control labour as flowing unproblematically from overall business strategies. On the other hand, writers adopting the strategic choice approach argued that managers were 'creative mediators' in shaping the outcomes of technological change in individual organisations. Managerial choices and decisions were influential in terms of: the reasons why new technology was introduced, in other words the objectives behind its introduction; choices over the actual implementation of change; and the organisational arrangements for the operation of the technology once it is implemented. This suggested a more complex and variable relationship between different management levels and functions in decision-making. This chapter will draw on recent empirical research to examine how managerial choice has influenced the process and outcome of change.

Management objectives in introducing new technology

Why do organisations adopt new technology? Is it, as 'innovation theorists' imply, simply a rational commercial calculation in response to technological imperatives or are managerial decisions, as many labour process writers argue, part of an overall strategy aimed at increasing management control by deskilling labour?

One of the more extensive pieces of case study research on the introduction of computing and information technology has been conducted by Buchanan and Boddy (1983). They carried out research in seven organisations in Scotland covering a variety of different industries, types of firm, and technology (see Figure 3.1). Their findings reveal that the management objectives behind change are the product of complex processes of strategic choice within organisations. The availability of new computing and information technology was a 'trigger' to, rather than determinant

Figure 3.1   Buchanan and Boddy's Scottish Case Studies

Case	Technology	Company
1	NC machine tools	Caterpillar Tractors
2	Computer co-ordinated measuring machines	Caterpillar Tractors
3	Computer-aided lofting	Govan Shipbuilders
4	Computer-aided lofting	Reach & Hall Architects
5	Word processors	Y-ARD marine engineering consultants
6	Computerised equipment controls	United Biscuits
7	Computerised process controls	Ciba–Geigy chemicals

of, procesess of managerial decision–making. Thus, decisions were not purely a matter of commercial calculation in response to technological imperatives but rather a product of political processes within organisations. Moreover, a mix of objectives was involved in decisions over how to use the new technology, and labour control objectives were not the decisive consideration in the strategic decisions made by senior managers when deciding to invest.

In each case studied the decision to adopt and choice of technology were 'championed' by key individuals or small groups who promoted the need to change, often in the context of opposition, scepticism and inertia on the part of other managers. The arguments of these 'promoters of change' were not only based on objective commercial analysis but were also designed to mobilise support, create coalitions with influential managers, and manage the flow of information to back their case. For example, in one instance – a study of the introduction of numerically controlled machine tools at a tractor plant in Glasgow – it was found that the idea of change had been initiated by middle managers in manufacturing-support functions who justified the investment in terms of productivity and cost benefits. They secured the agreement of senior manufacturing managers, who were concerned at the age of the existing plant, but had their proposals resisted by the plant accountants on the basis of a separate investment analysis and by factory managers who felt they would lose their control over the workflow (1983: 47–51). This points to the variety of management interests and objectives that may be involved in the process of technological change.

The various management objectives identified by Buchanan and Boddy in their case studies were classified into three main catergories. These were:

1 *Strategic objectives* concerned with external, market and customer-oriented goals such as improved product quality, expanding market share and maintaining market lead over competitors.
2 *Operating objectives* concerned with internal technical and performance-oriented goals such as improvements to product flexibility and reductions in plant running and labour costs.
3 *Control objectives* concerned with reducing uncertainty caused by reliance on informed human intervention in the control of work operations, for example by replacing humans with machines, improving management control over workflow, increasing the amount and speed at which performance information was made available to managers (Buchanan and Boddy 1983: 243–4; also Buchanan 1983: 75–6).

The pursuit of these objectives varied between management levels and functions. Senior managers tended to have strategic objectives. These goals were closely linked to the 'operating' objectives of middle line and financial managers. Control objectives did not figure significantly in either senior or middle managers' decisions to introduce new technology. There appeared, therefore, to be no overriding objective to introduce new technology in order to control labour. However, control objectives were to the fore in middle and junior line managers' thinking about how the new technology should be used once it was introduced (1983: 241–4). This point will be taken up below.

Buchanan and Boddy's findings are supported by a wide range of case study evidence (see for example, Francis *et al.* 1982; Jones 1982; McLoughlin 1983b; McLoughlin 1987; Wilkinson 1983; Willman and Winch 1985; Batstone and Gourlay 1986; Batstone *et al.* 1987; Clark *et al.* 1988). These studies also suggest that labour control objectives are, if they even figure at all, far from being the only factor in managers' decisions to adopt new technology, although this is not to deny that in some cases – in national newspapers for example – such considerations can exert a significant influence (see Martin 1984). Wilkinson in four case studies of the introduction of electronics-based technologies in batch engineering firms in the West Midlands[1] concluded that:

> Owners may sanction capital investment proposals which come from within organisations, and they may demand that certain levels of profitability are maintained; but this research provided no evidence that deskilling manpower policies were connected in any immediate sense to these demands.
>
> (1983: 97)

Similarly, Jones studied the introduction of NC machines (a technology cited as the 'classic case' of managements introducing new technology with the objective of deskilling labour by Braverman) in five batch engineering plants in south-west England and Wales. He found that labour costs were rarely, if at all, the decisive factor in decisions to invest in NC machines, while savings in operating costs and improvements in product quality were the principal determinants of managerial choices. He concluded:

> ...the evidence presented here further contradicts even modified theses about general and inherent tendencies to deskill because of 'laws' of capitalist exploitation and accumulation... This is not to deny that such motivations exist among manufacturing management. It merely reasserts that management calculation cannot be concerned solely with labour costs and utilisation.
>
> (Jones 1985: 198–9)

Available survey evidence appears to confirm this picture suggesting that a range of motives exists within the management function as a whole for the introduction of new technology, and that decision-making is distributed through different levels and functions. Surveys covering manufacturing firms and offices conducted for the Policy Studies Institute identified a variety of motivations for adopting new technology among the managerial respondents questioned (see Northcott *et al.* 1982; Bessant 1982; Steffens 1983).

More recent and comprehensive survey data confirms the manner in which managerial decision-making tends to be distributed through the different levels and functions according to the particular stage in the process of change concerned. For example, the WIRS found although there were significant variations between sectors. The general pattern in establishments forming part of larger organisations (eighty-four per cent of the sample), was for decisions to introduce change to be taken at higher levels in the enterprise. However, decisions over how change was to be introduced tended to be taken at establishment level (Daniel 1987: 86–7; see also Martin 1988).

To summarise: it appears that ideas that the introduction of new technology can be explained either in terms of 'rational' responses to commercial and technological imperatives or an overall management strategy aimed at increasing management control, are too simplistic. Case-study research on the reasons why new computing and information technologies have been introduced reveals that decisions are the outcomes of processes of strategic choice within organisations and that a variety of objectives may be involved. Moreover, the nature of the objectives pursued by managers varies according to level and function. Senior managers tend to be concerned with using new technology to improve the position of the organisation in the 'external' environment, middle managers with improving the 'internal' operating performance of the organisation, and lower-level line managers with the use of new technology to reduce informed human intervention and thereby increase managerial control of the work process. Motivations to use new technology to deskill labour and increase management control have rarely been decisive in senior managers' decisions to introduce new technology.

The regulation of labour and technological change

The relationship between senior managers' decisions to introduce new technology and policies concerned with the regulation of labour would appear to be more complex than implied in some versions of labour process theory. For example, while there is little evidence to suggest that issues such as labour costs and labour control are always decisive, it is important to recognise that the objectives behind decisions to adopt new technology do set parameters within which decisions over industrial relations and labour control policies are likely to be made. Thus, while such issues may not be an explicit component of overall business strategy, policy assumptions on such matters as labour costs can be implicit in senior managers' decision-making (Batstone *et al.* 1984; Child 1985).

This point appears to be borne out by the findings of a survey conducted by the Industrial Relations Research Unit at the University of Warwick which was concerned with the general impact of higher levels of management on workplace industrial relations (see Marginson *et al.* 1988). Part of the survey, which focused on the corporate and divisional levels of seventy-six firms, concerned the industrial relations aspects of decisions over technological change. The findings revealed that the great majority of corporate level managers did take industrial relations issues into account when deciding to introduce new technology, in particular the issues of levels of employment and work organisation (Martin 1988).

One of the difficulties with much existing case study and survey research is that, whilst pointing to the diversity and complexity surrounding decision-making over technological change, little is said about the way economic circumstances of the firm, or the type of innovations involved, might affect or constrain managerial aims and objectives with regard to the regulation of labour. Paul Willman (1986: 76–80; 1987) suggests a possible categorisation of the various factors which may set the parameters for managerial decision-making in this respect. The key factors are whether a firm's competitive position depends on maximising the performance, maximising the sales, or minimising the production costs of its products. Organisations operating in a relatively young product market are likely to seek to improve competitiveness by applying new technology to maximise the performance of their products, rather than seek to reduce the costs of production. These firms might be expected to introduce policies which aim to develop the skills and expertise of their workforce in order to improve quality and develop new products and services. This description would fit the science-based firms that – according to long wave theory – were the first to adopt microelectronics in new technology products (see Chapter 1).

In contrast, senior managers in firms in maturing product markets are likely to seek to maximise sales by using new technology to improve the efficiency and continuity of production. The implication for management policies for labour regulation here is that these might be directed at securing greater control over work, while at the same time attempting to secure employee acceptance of continuous technological change. Finally, senior managers in firms in mature product markets are likely to seek to maintain their competitive position by using new technology to reduce production costs. Where labour costs are the principal concern, they might be expected to develop policies which aim to improve labour productivity and utilisation, possibly involving a reduction and/or wholesale restructuring of the labour force. Significantly, suggests Willman, it is in mature industries concerned with cost minimisation that new process innovations based on microelectronics may have their greatest impact. This is supported by the current pattern of innovation which, as seen in Chapter 1, reveals that microelectronics-based innovations are now diffusing more generally into production processes across industry (Willman 1986: 168–79).

Clearly, then, there are likely to be variations in the implications of decisions to adopt new technology for managerial approaches to labour regulation, depending upon the nature of an organisation's product market, whether the change concerned is a product or process innovation, and if a process innovation, whether this is aimed at improving the quality of output or reducing costs. However, Willman cautions that it is not automatic that organisations introducing new technology in any of these circumstances will necessarily develop the 'appropriate' business strategy or approach to labour regulation. Rather, processes of 'strategic choice' may well be important mediators (1986: 77). Similarly, Batstone *et al.* (1984) point to the importance of strategic choices in mediating the link between overall business strategy and labour regulation. In this sense the overall business strategy developed by an organisation in response to changes in the commercial and technical environment might best be seen as a 'corporate steering device' (Child 1985) which sets the parameters within which sub-groups of managers develop policies and approaches to implementing technological change.

Child (1984; 1985) has provisionally identified four choices or 'policy directions' with respect to labour regulation which may be applicable to organisations introducing new computing and information technologies into their production processes. These are:

1 Elimination of direct labour by substituting new technology for human involvement in production to provide completely automated continuous production processes.

2 Sub-contracting of work outside the organisation by using new technology to enable professional and managerial staff to work on a contract basis from home using computer workstations linked to company headquarters. Similar developments may be possible where the high reliability of equipment allows maintenance work to be sub-contracted.

3 Improved labour flexibility ('polyvalence') by breaking down existing horizontal and vertical skill demarcations between workers, thus increasing the range of tasks that they can perform and/or the amount of discretion that they can exercise over task-performance.

4 Degradation of jobs by using new technology to deskill tasks and increase direct management control over work in line with the Taylorist practices identified by Braverman.

In practice, the appropriate balance between these options, suggests Child, is likely to vary according to the circumstances of particular organisations, such as product and labour market conditions and the nature of the production task.

As Child also notes, such policies have immediate implications for what labour economists refer to as an organisation's 'internal labour market'. Researchers at the Institute of Manpower Studies at Sussex University have described the possibilities for the restructuring of internal labour markets in terms of a model of the 'flexible firm' (see Figure 3.2) (Atkinson 1984; Atkinson and Meager 1986; Preece 1986). This theoretical model distinguishes between the 'core', 'peripheral' and 'external' components of the organisation's workforce. 'Core' workers are those whose skills are most essential to the conduct of the main activities of the firm. They enjoy primary labour market status which is reflected in relatively high job security, pay and status. They are required to accept 'functional flexibility', so that changes to their jobs (for example the erosion of horizontal and vertical demarcations with other skill groups) can easily be made when required. 'Peripheral workers' conduct less critical activities and are hired on employment contracts (temporary, part-time, flexible working hours, etc.) which allow for 'numerical flexibility' by the easy adjustment of their numbers as the scale of operations changes. 'External workers' are those whose activities are 'distanced' from the firm in the sense that they are no longer employed by it, their employment contract in effect being substituted by a commercial contract with another employer (for a critique of this model see Pollert 1987).

The 'policy directions' identified by Child suggest that new technology may enable managers to bring about reductions in overall employment levels and increase 'numerical flexibility' among peripheral workers by eliminating and deskilling jobs. It may also facilitate reductions in employment levels and increases in the 'external' workforce by allowing some 'core' and 'peripheral' work to be sub-contracted. Finally, the erosion of skill demarcations when new technology is introduced suggests an increase in 'functional flexibility' amongst 'core workers'. In short, the introduction of new computing and information technology may enable managers to develop labour regulation policies which result in a radical restructuring of an organisation's labour force and be one factor contributing to more flexible forms of work and employment. To what extent though have managers adopted such innovations in their approach to regulating labour when new technology has been introduced?

Figure 3.2   The flexible firm

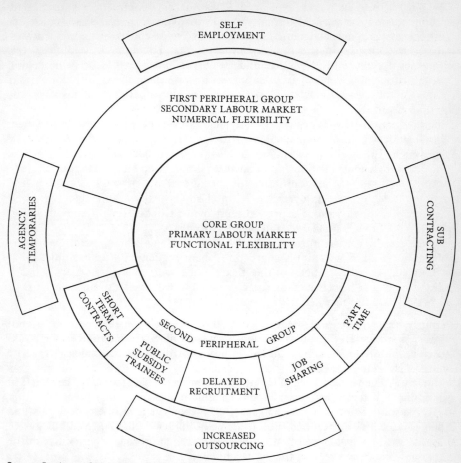

SELF
EMPLOYMENT

FIRST PERIPHERAL GROUP
SECONDARY LABOUR MARKET
NUMERICAL FLEXIBILITY

AGENCY
TEMPORARIES

SUB
CONTRACTING

CORE GROUP
PRIMARY LABOUR MARKET
FUNCTIONAL FLEXIBILITY

SHORT
TERM
CONTRACTS

PUBLIC
SUBSIDY
TRAINEES

SECOND   PERIPHERAL   GROUP

PART
TIME

JOB
SHARING

DELAYED
RECRUITMENT

INCREASED
OUTSOURCING

Source: Institute of Manpower Studies (1984)

Innovations in labour regulation and the role of personnel management

Some commentators have argued that during the recession the introduction of new technology has occasioned the emergence of a 'new management strategy' for handling the industrial relations issues raised by change. This strategy involves a firmer line being taken with trade unions and a willingness to consult and involve employees only within tightly defined limits (see Northcott and Rogers 1985: 36–8). However, it may be misleading to see such strategies as 'innovative' approaches to labour regulation in the context of the introduction of new computing and information technologies. On the one hand, there is evidence from industries such as automobiles and national newspapers, that some managers have taken a tougher line with trade unions, and have seen direct communication with the workforce as a means of gaining the acceptance of change in situations where the labour market position of unions is weak (see for examples, Willman and Winch 1985; Martin 1984). On the other hand, it is by no means clear that such 'new strategies' represent anything more than an attempt to reassert managerial prerogatives based on traditional assumptions and values. For example, Davies concluded from a study of technological change in the brewing industry that managers viewed the weak position of unions as an advantage and an opportunity to avoid bargaining over change or disclosing detailed information on management plans. The assumption was that unions would react negatively and that the superiority of management expertise meant that the shop floor had little to contribute (1986: 181–3).

Sheila Rothwell (1984) studied twenty-one firms engaged in introducing new technology drawn from both manufacturing and service sectors (see Figure 3.3). According to Rothwell, an organisation's 'employment policies' are likely to vary along five dimensions: relationship to business strategy; extent of a long-term future planning orientation; the significance of the role of the personnel function; the extent to which an employee-centred philosophy exists; and the degree of consistency between various aspects of employment policy (manpower planning, industrial relations, pay) and overall business strategy.

To the extent that the introduction of new technology is being accompanied by innovations in an organisation's approch to labour regulation – for example, along the lines suggested by notions of the 'flexible firm' – it might be expected that a 'high' score would be achieved on some if not all of Rothwell's five dimensions. In fact her case studies revealed that:

1 Employment policies tended to be a 'lower-order' consideration in overall business strategies. Decisions to adopt new technology tended to overlook issues such as the need for training courses to be completed if the technology was to be used effectively.
2 There was a general absence of long-term planning to take account of the effects of adopting new technology on manpower requirements.
3 The personnel function was not usually involved in investment decisions and played a reactive rather than proactive role in implementing new technology.
4 There was little evidence of an employee-centred approach, in particular in the area of work design around the new technology.
5 Various aspects of employment policy did show some changes in approach towards internal labour market structure in a few organisations.

Rothwell concluded that the general lack of change in employment policies at a strategic level probably reflected managerial preferences 'to pour new wine into old bottles' in order to reduce uncertainty, in particular by avoiding linking techno-logical change to wholesale changes in the approach to employment and industrial relations. As one manager commented: 'If I can get away with changing three things, rather than six, I'll do that, because people can't take too much' (1984: 111).

The typicality of these findings can be explored further by looking at more detailed evidence on the role of personnel and industrial specialists in the process of technological change. Preece and Harrison (1986) argue that the major implications of new technology for work and organisation might enhance the role of personnel and industrial relations managers as 'would-be change agents' in developing new approaches to labour regulation. In other words, personnel and industrial relations specialists might be expected to be called upon, in what Thomason (1981) refers to as their 'organisational consultant' role as change increasingly involves dealing with 'human problems' requiring specialist expertise. On the other hand, it can be argued

Figure 3.3 Summary of Rothwell's case studies

Industry/case	Application of new technology	Employees in case unit	Trade unions recognised
Engineering			
Electronics	Materials and production control	590	✗
Heaters	Manufacturing	450	✓
Print	Manufacturing	200	✓
Engines (2)	Assembly working	1200	✓
	Manufacturing	670	✓
Power tools	Materials and production control	1400	✗
Food, drink & tobacco			
Confections (2)	Chocolate bar manufacturing	1800	✓
Packet food	Automated warehousing and packing	75	✓
Liquid food (2)	Quality control	330	✓
	Accounting	70	✓
Chemicals and Pharmaceuticals			
Photo products (2)	Materials and production control	250	✓
	Order processing	160	✓
Tablets	Quality control	300	✓
Infusions	Materials and production control	900	✗
Allergens	Materials and production control	450	✓
Drugs	Materials and production control	2800	✓
Wholesale distribution			
Agent	Stock control and order processing	111	✓
Wholesaler	Order processing	300	✓
Own distributor	Stock control	3500	✓
Mail order	Automated warehousing	1200	✓
Stationer	Order processing	3500	✓
Finance & banking			
Insurance	Order processing/ accounting	500	✓
Computing	Sales & service of computers	60	✗
Public utility			
Utility	Customer service/ order processing	4000	✓

Source: Rothwell (1984)

that the historical development of personnel and industrial relations management in Britain has tended to emphasise a reactive rather than proactive approach (see Watson 1977; 1986: 172–209) concentrating on 'fire fighting' and *ad hoc* decision-making rather than strategic thinking (Purcell and Sisson 1983). This suggests that personnel specialists are marginal to managerial decisions and planning over technological change. This rather than being involved in the initial stages of change, personnel and industrial relations managers will be called upon only at subsequent stages to deal with employment and control problems once they arise.

The role played by personnel specialists in seven manufacturing firms in the Midlands and North of England was studied by Preece and Harrison (1986). The cases illustrated the varied nature of the personnel involvement in technology-related changes and the widely varying contribution that they could make. These ranged from cases where personnel managers played an important and influential role supporting line-management decision-making, to situations where there was no personnel involvement at all and where expert involvement might have avoided costly mistakes being made by line-managers.

For example, in the case of an American-owned engineering company, the plant of around 500 employees had a well-developed personnel function. The plant was first occupied in 1972 as a 'greenfield' site and had subsequently introduced a range of CAD and CAM technologies. This created an atmosphere in which technological change was viewed as a 'way of life' and established structures and procedures were developed to handle the introduction and operation of new systems. The role of personnel in the process of change itself was deliberately 'low profile' in order to retain an air of normality when new technology was introduced. However, personnel had an important background influence in advising line managers and attending to any employment and control issues that arose – for example, pay, grading and training.

This situation can be contrasted to that in the plant of a specialist truck body builder with 220 employees. The personnel function in this plant consisted of one personnel manager nearing retirement whose role was essentially administrative. One consequence was that the company had little idea of the skills and knowledge profile of its workforce. When the company installed its first new technology in the form of NC and CNC machine tools the personnel manager had no involvement whatsoever. One consequence was that the absence among the workforce of electronic and programming skills to operate the machines was not realised by line managers until after the machines were installed. Even at this stage there was no attempt to involve personnel. Instead the machines stood idle until an external consultant could be employed to produce the required computer programs.

Rothwell's studies, cited above, point to more involvement on the part of personnel than in this latter case. She found that although it was rare for senior personnel directors to have any major involvement in feasibility studies initiating change or leading to the decision to adopt a particular technology, the personnel function was more involved at subsequent stages of change (1984: 44).

This finding is supported by Clegg and Kemp (1986). They studied the introduction of a 'state of the art' FMS system in a large plant of a UK-based manufacturing company, the planning process for which lasted several years. They describe this process as illustrative of the 'sequential method' of introducing new technology where system design decisions are taken first, after which the human implications of these decisions are considered. In the case concerned, personnel managers were excluded from the system design stage and were subsequently required to react to the technical decisions that had already been made. The effect was that personnel and industrial relations issues were 'squeezed out' at the design stage, subsequent choices over human aspects were constrained by prior technical decisions, and no coherent strategic plan concerning the human aspects of the new system was developed. As a result of the lack of input from personnel, not only were the human aspects of operating the system considered only at a late stage in the process of change, but when they were considered decisions were fragmented and did not constitute a coherent policy (1986: 9).

How typical though are these case-study findings? Is the general pattern for personnel to be highly influential at all stages of change, involved only after major decisions have been taken, or simply not involved at all? The Workplace Industrial Relations Survey (WIRS) provides more wide-ranging evidence from organisations in manufacturing which allows a more general picture to be constructed. This tends to suggest that the lack of involvement of the personnel manager in the truck plant studied by Preece and Harrison may well be the most typical. First, it was found that only twenty per cent of manufacturing establishments had personnel managers, although these tended to be the largest workplaces in terms of numbers of employees. Second, as Table 3.1 indicates, in forty-six per cent of cases personnel managers were not involved at all in the introduction of new microelectronics technologies (denoted in the Table as 'advanced technical change'). Third, where personnel was involved, in only fifteen per cent of cases was this in the initial

Table 3.1 Works managers' accounts of the role of the personnel department in manufacturing establishments in the introduction of major change[2]

Column percentages[a]

	Total	*Advanced technical change*	*Conventional technical change*	*Organisational change*
Personnel department involved	46	50	(13)[b]	(80)
Personnel not involved	52	46	(87)	(20)
Not stated	2	4	(1)	–
Stage of involvement				
Decision to change	14	15	(1)	(30)
Immediately after decision to change	20	19	(2)	(50)
After decision to tell workers	6	9	(2)	–
Later stage	6	7	(8)	–
Base: works managers who reported a major change in the previous three years				
Unweighted	*241*	*176*	*40*	*25*
Weighted	*56*	*37*	*12*	*7*

[a] See note C
[b] See note B

Source: Daniel (1987)

decision to adopt new technology, in most cases involvement coming in the later stages of change – in other words as part of a process of 'sequential' decision–making as described by Clegg and Kemp.

These findings are supported by those of the Warwick survey referred to above. Here no evidence was found to suggest that the introduction of new technology was directly related to an increase in the role of the personnel function, although there was a tendency for the influence of personnel to have grown where managers had engaged in discussions over the introduction of new technology with union representatives (Martin, 1988).

The lack of involvement of personnel managers is made even more striking when compared to their role in organisational changes without a technological content (see Table 3.1). The WIRS found that in manufacturing personnel was involved in eighty per cent of such cases, and that this involvement was nearly always from the decision to change or immediately after (Daniel 1987: 107–9). Comparing these findings with those of earlier studies of technological change Daniel concludes: 'it appears that very little has altered since the early 1960s, and that technical change is still largely seen as a technical matter within which there is no established role or function for personnel management' (1987: 110). What is surprising, as Daniel notes, is that there is clear evidence that the full and early involvement of personnel in technological change, where it does occur, helps to promote worker acceptance of change. Indeed, the survey found a strong and consistent relationship between the early involvement of personnel managers and favourable worker reactions to advanced technological change (1987: 110).

To summarise so far. It appears that policies for the management of labour do not flow unproblematically from overall business strategies as suggested by many labour process writers. Although business strategies may set the parameters within which such policies may be developed when new technology is introduced, for example by specifying a general requirement to reduce labour costs, they do not necessarily specify the policies to be pursued. To this extent there is strategic choice within any organisation over the precise kind of labour management policy to accompany technological change.

What is significant about the findings of recent research is the lack of innovation in approaches to the regulation of labour when new technology is introduced. In other words, although strategic choices may exist which suggest possibilities for a radical restructuring of work and employment within organisations, there is little evidence that managements have in general actively sought to take such decisions and develop 'new strategies' for labour regulation. Perhaps the clearest indication of this is the marginal role played by personnel specialists who might have been expected to play a leading role as 'organisational consultants' if new policies were to be developed at a strategic level. However, if anything, evidence confirms that they tend to become involved only after decisions to adopt new technology have been taken and in response to discussions with trade unions. In other words, their role is marginal and reactive rather than central and proactive. This brings us to the choices faced by managers with regard to the implementation of new technology – in particular the issues of how far a participatory approach should be adopted and the form of management organisation that should be used to manage the change.

Management choice and implementing new technology

It was suggested above that business strategies leading to decisions to adopt new technology set the parameters within which lower-level managers responsible for working out detailed policies have to operate. This raises the important question of how far decisions at a senior level are actually carried through in the choices made by managers at the workplace during subsequent stages in the process of change. This is referred to by Child (1985) as the extent of 'attenuation' or the 'tightness of coupling' between corporate strategy and the actions, or as we have termed them elsewhere 'sub-strategies' (see McLoughlin *et al.* 1985), of lower-level managers.

Whilst the 'sub-strategies' behind the choices made by lower-level managers may fulfil the role of 'plugging gaps' left by corporate decisions to introduce new technology, it is equally possible that managerial actions, or a lack of them, at subsequent stages of change may lead to consequences not anticipated or intended at senior levels in the organisation. In this sense, the actions of middle and junior managers can in certain circumstances be regarded as a 'subversive activity' to the extent that they may actually frustrate corporate objectives (Buchanan 1986: 77). This section will explore the significance of the actions of middle and junior managers at the implementation stage of change. The following section examines management sub-strategies in using new equipment or systems once they are operational.

Management sub-strategies for implementing new technology

It has already been noted that decisions to adopt new technology open up or trigger further management choices at subsequent stages in the process of change. Having decided to adopt new technology and selected or designed the required equipment, managers then face decisions as to how to implement the new equipment and systems at workplace level. In principle, choices range along two dimensions (see Francis 1986: 171–96; Davies 1986: 66–72; Child 1984: 268–93 for discussion). Firstly, there is a question of how far a participatory or non-participatory approach is taken in relation to the workforce. At one extreme managers might take the view that all decisions regarding the introduction of new technology are a managerial prerogative and that a minimum of information should be made available to employees. Alternatively, they may seek to communicate and consult fully with employees about their decision to introduce new technology. Finally, in unionised environments managers have a choice as to whether technological change should be an issue for joint consultation and/or collective bargaining. Obviously, as will be seen in Chapter 4, trade unions may well have views of their own on the extent to which managerial decisions should be subject to joint regulation.

Secondly, there is a question of how managers should organise themselves to manage the change, whether decisions should be made from the 'top-down ' or whether a 'bottom-up' approach should be adopted. The 'top-down' approach implies that change is managed centrally by a project team, task force or individual manager. One advantage of this approach is that it enables a high degree of senior management control and maintains continuity of management responsibility throughout the process of change. A disadvantage is that the middle and junior line-managers who will be the 'end-users' of the technology may have little influence over the way change is implemented. A 'bottom-up' approach involves delegating responsibility for managing change to the line-managers in the individual plants, areas of operation, or functional departments who will ultimately be responsible for using the new technology. One obvious advantage is that the requirements of the end-users are more likely to be met since they will be responsible for making decisions at critical junctures during implementation. A disadvantage is that senior managers will have less direct control over the precise way change is managed, and continuity may be more difficult to maintain, especially where line-managers give a higher priority to their day-to-day responsibilities than to the management of change. The range of choices open to managers in developing sub-strategies is summarised in Figure 3.4.

Organisation theorists who have adopted a contingency approach suggest that there is no 'one best way' to manage change, although nearly all academic commentators point to the advantages of adopting some form of participatory approach towards the

Figure 3.4   Management choices in implementing new technology

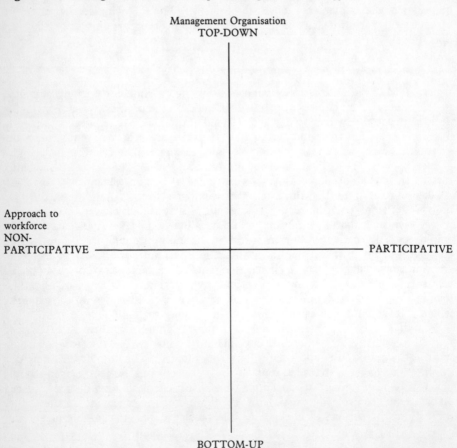

workforce. 'Situational factors' which may influence management choices include the timescale of the project; the nature and strength of possible sources of resistance; the identity and influence of 'promoters of change'; existing collective bargaining arrangements; and the nature of trade union reactions. The more a technological change is important to organisational performance and commercial survival the more managers are likely to adopt a top–down approach. The more strong resistance is anticipated from a trade union within an established collective bargaining framework, the more likely a bottom–up approach may be adopted (see Davies 1986: 70).

An important point is that the 'sub-strategy' adopted by managers in implementing new technology is likely to be the outcome of choice and negotiation rather than merely, as contingency theory implies, a rational selection from alternatives within a given situation (see also Davies 1986: 71). Some of these points will now be illustrated by reference to three case studies from research in which the authors have been involved. Each case involves a different implementation sub-strategy which emerged as a result of choice and negotiation within the organisation. As will be seen, the choices made had profound implications for the successful achievement of corporate objectives, in some cases acting to facilitate and in others to frustrate.

Three cases of implementation

The first example is provided by British Rail and their attempt to introduce a computer-based freight information system ('TOPS') into the rail network during the 1970s (see McLoughlin *et al.* 1983; 1985). The case provides an excellent illustration of an implementation sub-strategy involving a combination of top-down decision-making and a mixture of a participatory and authoritarian approach at different levels within the organisation. It was designed to overcome potential management and trade union resistance to change by securing a 'tight coupling' between corporate objectives and the manner in which they were implemented at workplace level.

The decision to invest in new technology was based on 'strategic objectives', concerned with arresting the severe economic difficulties of the freight business in the face of declining traffic, 'operating objectives', concerned with improving the efficiency of freight operations, and 'control objectives', aimed at improving the utilisation of freight rolling stock and locomotives. The sum invested was in the region of £13 million (1971 prices) and was British Rail's first venture into large-scale, on-line computer information systems in the control of rolling stock and train movements. The nature of the innovation, and the economic circumstances of its introduction, made the change fraught with risks of failure, especially as a particularly short timescale for implementation was approved as part of the investment justification.

The TOPS project had been 'championed' by British Rail's Chief Executive who acted as a 'promoter of change'. Nevertheless, there had also been considerable trepidation among other Board members and senior managers that implementation would be frustrated and delayed by either failings in the technology or a combination of management and trade union resistance to change – a fate which had befallen attempts at innovation and change in the organisation in the past. In particular there was concern over potential resistance to the new technology on the part of middle and junior line-managers in the operating regions. In the past much of their autonomy and authority had rested on their ability to manipulate and control the flow of information to senior management. The TOPS system, among other things, promised to provide senior managers with rapid access to up-to-date and accurate information which would make the performance of line-managers far more 'visible' (for more details see Chapter 7).

In response, the Chief Executive developed a 'top-down' approach to managing implementation. This involved the creation of a specialist multi-disciplinary 'task force' reporting direct to him and autonomous from both headquarters functions and the regional management. The task force was responsible for handling all aspects of implementation, in particular the installation of equipment and the training of staff in a 'rolling' programme across the network. Complete authority was invested in the task force's Project Manager, which not only allowed the task force to override any local management resistance, but also meant he played a leading role in discussions with the railway trade unions. Significantly, the task force did not seek to bypass the existing national industrial relations framework, but made a deliberate and conscious effort to provide national union leaders with information on their plans from the earliest stage. However, while adopting this participatory approach, managers were also adamant that the change should not spark off national claims for increased payments to use the technology.

Discussions with the unions at national level were kept within the industry's procedures for joint consultation and were not allowed by senior managers to become the subject of national negotiations. In the event, national trade union leaders gave the system positive backing: first, because unless something was done to arrest the decline of the freight business then large job losses would have been inevitable; second, because the TOPS system did not involve job losses in manual grades and actually created new jobs for white-collar grades; third, the changes in working practices required provided possibility for negotiating extra payments for their members, at least at local level; fourth, there were no plans to use TOPS to monitor the productivity of train crews, a 'Big Brother' connotation of the system which worried the train drivers' union, ASLEF.

From management's viewpoint the success of the approach was illustrated by the fact that the TOPS system was installed on time and within budget, previously almost unheard of within the industry, and improvements in operational efficiency were achieved almost immediately. Middle and lower management resistance to change was either ignored or challenged by the project task force, and willingness to consult with the unions from the earliest stages was repaid by the active involvement of national leaders in quickly resolving local disputes which arose during implementation. However, despite this the task force approach itself was not considered a success by senior managers at headquarters or in the regions. Considerable animosity had been created by the 'top-down' authoritarian way change had been imposed on middle and junior line-managers, and it was considered politically expedient to disband the task force once implementation was complete, although several enhancements to the system were planned and changes to both supervisory and management organisation still had to be accomplished. It is interesting to note that subsequent technological changes within BR, including the Advanced Passenger Train, have not employed the task force approach, resulting in the latter case in the development of management resistance which ultimately became a key factor in the failure of the project (see Potter 1987).

The second example is drawn from British Telecom and their introduction of a semi-electronic telephone exchange (TXE4) to replace the electromechanical 'Strowger' system during the late 1970s and early 1980s (see Clark *et al.* 1988). This case provides an excellent illustration of an approach to implementation, involving a mixture of top-down and bottom-up sub-strategies, which resulted in the 'attenuation' of corporate objectives. The decision to introduce the TXE4 exchanges had been taken in the early 1970s by the Post Office, at that time responsible for the public telephone service as well as the postal service. The decision was based on 'strategic objectives', aimed at meeting the growing demand for telecommunications services and improving the quality of the service but these were linked to 'operating objectives', directed at reducing exchange maintenance costs, and 'control objectives' aimed at cutting staff levels and improving the productivity of maintenance technicians.

The corporate strategy behind modernisation also reflected the problematic nature of the relationship between business strategy and approaches to labour regulation in the post and telecommunications businesses at the time (see Batstone *et al.* 1984 for discussion). For example, the industrial relations policy developed by senior management provided the basis for an agreed national industrial relations framework to govern the introduction of new technology with the trade unions. However, this framework and the agreements which underpinned it (covering job security and minimum staffing levels) were based on traditional patterns of labour relations in the industry that did not correspond with the new commercial emphasis in overall business strategy. Moreover, the agreements themselves also left open to question and interpretation how their provisions were to be implemented at workplace level. For example, whilst laying down an apparently precise formula by which new staffing levels could be decided, in practice the 'small print' of the agreement concerned left considerable room – or 'design space' – for further interpretation and negotiation at local level (see Clark *et al.* 1988: 71–5). In similar fashion, senior management's 'control objectives' initially provided little more than national policy and performance guidelines which left open the question of how they were to be achieved in practice, in particular how maintenance work was to be organised and supervised at local level (ibid. 41–50).

While the selection of exchanges for modernisation was planned centrally, the approach taken by British Telecom was to delegate responsibility for actual implementation to middle line-management in each telephone area. Other than the policy guidelines and national agreements already mentioned, area managements had little further guidance on how to set about their task. In other words, they had considerable 'design space' within which 'sub-strategies' to implement the new technology could be developed. This can be illustrated by looking at the sub-strategies developed by three separate area managements (ibid. 50–4, 76–81). For example, responsibility for implementing the new technology was given different priorities in each of the three areas, and in only one case was one manager given overall responsibility. Similarly, local managers usually interpreted the national agreement on staffing levels generously in order to keep staffing in individual exchanges high enough to cope with absences through training, unexpected maintenance problems, and staffing requirements in the area as a whole.

Finally, on issues such as selection of staff and the organisation and supervision of work, where no national policy or guidelines existed, area managers had the opportunity to develop their own solutions. On the issue of staff selection, two area managements negotiated an informal agreement with local union representatives on the criteria to be employed, while the other decided this unilaterally. One effect of these actions in all three cases was to contribute to the exclusion of junior technicians from selection for retraining – something never expressly intended by senior managers. The general issue of the control and supervision of work is examined in some detail in Chapter 7 and the outcomes in this case are used as an example there. Suffice it to say that variations in middle and junior line-managers' responses again led to outcomes not anticipated or intended at corporate level.

The final case concerns the introduction of a new microelectronics–based light-weight TV camera, developed in the United States for electronic news gathering (ENG) and adopted in Britain in the early 1980s by a regional independent TV company (see Jacobs 1983; Clark *et al*. 1984; McLoughlin *et al*. 1985). This case is similar to the BR example in so far as senior management delegated responsibility for managing implementation to a single authority. However, whilst a 'top-down' approach was evident in this sense, the ultimate solution to the principal problem which management confronted during implementation came in the form of an almost purely 'bottom-up' approach. This occurred because management were prepared to go beyond the requirements of the national industrial relations framework covering the industry to allow the development of joint procedures which ensured a high degree of involvement in the decision-making process on the part of the union concerned (the Association of Cinematograph and Television Technicians, ACTT) and the technician workforce. In this sense the case differs from both the BR and BT examples.

The company concerned had recently been awarded the franchise for a local broadcasting region by the Independent Broadcasting Authority. The Company's proposal for the franchise included a strategic commitment to introduce ENG technology as part of its plans for a new style local news programme. Traditionally news pictures and features had been single-camera productions shot and recorded on film by specialist technicians. Unlike film, ENG cameras use video recording technology which eliminates the need for film processing and speeds up the time it takes to bring pictures from the scene to the viewer. With the aid of appropriate telecommunications links it is possible to beam 'live' pictures directly onto the screen. The news programme 'product' proposed by the company was intended to be sharp, immediate and full of 'hard news' items. This style of programme could be produced only by taking advantage of operational improvements enabled by the technical capabilities of ENG technology. In similar fashion to BR, the Company's corporate strategy involved tight time and cost constraints for the introduction of the technology. In particular, since the news programme was regarded as the regional TV station's 'flagship' product, it was particularly keen to have ENG equipment implemented in time for the commencement of broadcasting. This meant ENG had to be introduced within a twelve-month period.

The principal implementation problems facing the Company were industrial relations issues. Since the job-satisfaction of TV camera staff as an occupational group was derived in part from working with the latest technology, the problem was not one of workforce resistance to change as such. Rather, difficulties arose in relation to who would use the ENG cameras since two separate groups of television technicians – 'video' (concerned with directed multi-camera work in the studio and outside broadcasts) and 'film' (concerned with news pictures and other single-camera location work) – laid claim to the new work. On the one hand, there were management control advantages in allocating the new work to video technicians in terms of the lower cost and greater flexibility of their labour. On the other hand, allocating the work to the film technicians carried benefits in terms of their traditional skills and experience. Moreover, the Company was also bound by a nationally agreed code of practice with the ITV industry trade unions on the introduction of new technology, which included a procedural commitment to the fullest local consultation and negotiation. In addition, the employers and the ACTT had made a national commitment to the speedy introduction of ENG.

The approach taken by senior managers was in fact to demonstrate a clear commitment to participation, but under the control of a well-defined central management authority. At the start of the process of change, information on corporate strategy was given directly by senior board members to a meeting of the whole workforce. This was followed up by regular joint consultation with local ACTT representatives as well as direct consultation with members of the workforce concerned, and also feedback with the workforce through the preparation and wide distribution of discussion papers on management strategy. Senior board members delegated responsibility for introducing ENG to the general manager of the site concerned. He was well known to the workforce but was not identified with either the video or film department. In particular his personal style as a quick-talking, astute, and hard negotiator meant he enjoyed the trust of union officials.

The extent of management's commitment to participation is evidenced by the fact that over a five-month period the general manager was involved in over sixty formal and informal meetings with the trade union and the departments likely to be affected. In this way management kept in touch with the views of the union and the workforce and also kept the workforce and union directly informed of management's thinking. Given this participative approach it is also relevant to consider the union's response. In an attempt to overcome the problem of potential divisions among its own members an internal union sub-committee charged with developing a positive and united union strategy towards ENG was formed. At the same time, in order to prioritise the issues raised by the new technology, the union persuaded management to free its two main lay officers from normal duties for the duration of the negotiations. Finally, in order to maximise their expertise and experience, a group of past shop stewards were co-opted on to the executive committee of the local union branch.

Subsequent negotiations were dominated by the question of which group of technicians should be allocated the new ENG work. Management's initial proposals were presented to the workforce in a series of discussion papers which suggested that ENG work would normally be allocated to film technicians. These proposals were unacceptable to the video technicians. In the meantime the special union sub-committee proposed the creation of a new integrated grade of 'television technician' drawing on members of both film and video departments. This proposal was unacceptable to the film technicians. However, rather than exacerbating divisions within the workforce, the rehearsal of these proposals within a participative framework actually paved the way for agreement. This encouraged the union sub-committee and the camera technicians to continue seeking solutions to the problem, a process actively encouraged by management who, in the interests of 'good industrial relations', made it clear that a solution devised by the union which had the backing of both groups of technicians, and was not excessively costly, would be accepted.

In the event this is exactly what happened. A detailed solution was worked out and, after discussion with the affected groups of technicians, was put by the union to management who accepted the proposal. This involved a 50/50 split of ENG work between the film and video departments. Ironically, there was then a considerable delay before the technology was actually used. Once negotiations had been successfully concluded a change in the financial position of the company meant that they were no longer able to purchase the equipment as quickly as they had intended.

In sum, these three cases illustrate different approaches to the problems of implementing technological change. In the BR case this involved a top-down/ authoritarian approach based around a project task force and at the same time a participatory commitment to consult with the trade unions within the existing industrial relations framework of the industry. The BT case involved a mixture of top-down and bottom-up approaches with a commitment to a participatory approach with the trade unions at national level and where many aspects of managing implementation were delegated to local operating areas. A negotiated framework for implementing the technology was agreed with the unions at national level but this still left considerable room for further interpretation and negotiation at local level. The ENG case involved a top-down/participatory approach where management responsibility was delegated to a single manager rather than task force. In this case the extent of participation was far more extensive than in the BT and BR cases, going beyond an already participative framework established at national level to develop additional mechanisms for involvement within the company itself. Participation and involvement in this case did not mean simply consulting with trade unions, but also regular meetings with individuals and groups of employees likely to be affected directly by the change.

The British Telecom case provides an example of an approach to implementation which resulted in a very 'loose coupling' between corporate objectives and the way the strategy to introduce new technology was implemented at workplace level. It contrasts with the British Rail and ENG cases in that more 'design space' was left for middle- and lower-level managers by senior management's strategy towards introducing the technology, especially in relation to industrial relations. Whilst the BR and ENG cases illustrated 'top-down' solutions applied at all critical junctures during implementation, the BT approach tended to result in 'bottom-up' solutions which varied from area to area.

All three approaches involved benefits and costs. The BR approach had the benefit of ensuring high-level management control and continuity, but with the cost that change was to a large extent imposed on middle and junior managers. The TV Company's approach also ensured high levels of management control and continuity, but the commitment to participation meant that a compromise solution to the work allocation problem had to be accepted in the interests of 'good industrial relations'. The BT approach had the benefit of allowing managers ultimately responsible for using the new technology to develop their own solutions to problems and issues, but with the cost that lower-level management's responses might be variable in their effectiveness, thereby increasing the risk of unintended consequences arising from their actions. All three cases illustrate the importance of managerial decision-making at critical junctures in the implementation stage of change if new technology is to be introduced successfully.

Management choice and the operation of new technology

Decisions to adopt new technology not only trigger choices with regard to implementation but also raise organisational issues of how the technology is to be used once operational. In particular, questions are raised about the nature of job content, the pattern of work organisation, and how work is to be supervised and controlled. According to Buchanan, when new technology is introduced, 'the key decisions that affect organisational performance are those concerning the reorganisation of work that accompanies technical change' (1986: 78). As will be seen in subsequent chapters, management choice has a critical bearing on these 'key decisions'.

A predominant view in academic writing in recent years – in particular that influenced by labour process theory – has been that the assumptions underlying management decision-making over the organisation and control of work have been dominated by a relatively narrow set of criteria. These, it is claimed, are based on the theories of scientific management, in particular as influenced at the turn of the century by its most well-known exponent, F. W. Taylor.

For Braverman, 'Taylorism' was synonymous with the theory and practice of capitalist management *in toto* (see Chapter 2). However, other writers influenced by labour process theory have attempted to provide a more differentiated and sophisticated analysis of the influence of Taylorist ideas on management practice. For example, Littler (1982; 1985; Littler and Salaman 1984) distinguishes three different areas in which management practices based on these ideas can be implemented: work design; management control structures; and the employment relationship. Taylorist practices in work design seek to deskill jobs through a process of task fragmentation. According to Littler, this fragmentation process involves a progressive reduction in the task range of individual jobs, a reduction in the discretionary components of each job, and a reduction in the skill requirements of any task in order to minimise job-learning times (1985: 11). Taylorist practices in the design of management structures seek to reduce worker autonomy by increasing management control and supervision of work performance, to break workgroup solidarity, and to motivate employees through individual incentives (1982: 52–5). Finally, implicit in these changes is a fundamental transformation of the employment relationship within the enterprise which seeks to maximise employee substitutability and to minimise the organisation's dependence on individuals' skills and motivation. Littler does note, however, that such 'hire-and-fire' employment relations can give way to more paternalistic relationships (Littler 1982: 55–7).

Littler also argues that Taylor's ideas have had a varied influence in different national contexts, both in terms of the extent, timing, and nature of their diffusion and the degree to which they have become established as an ideology of management practice. The predominance of Taylor's ideas has also been affected by the extent to which other versions of 'scientific management', such as Henry Ford's ideas on continuous-flow assembly-lines, have been influential (see Littler 1982; 1985; Littler and Salaman 1984). In the case of Britain, Littler has argued that, whilst

Taylor's ideas have never been widely accepted as an ideology by senior managers, they – or rather a derivative version of 'scientific management' known as the Bedaux system[3] – have had a significant effect on management practice at workplace level. This, according to Littler, has meant that 'in general the direct and indirect influence of Taylorism on factory jobs has been extensive', such that work-design and technology-design have become 'imbued with a neo-Taylorism' (Littler 1986: 13).

However, other writers have argued that there are serious grounds for doubting Littler's albeit qualified assertions regarding the widespread influence of neo-Taylorism on management practices. For example, Batstone *et al.* (1987) argue that the Bedaux system's introduction in Britain was hardly as decisive a development in labour control as Littler appears to believe. They point out that his case rests on the fact that at some point in the period between the two world wars 245 firms adopted this system. As they argue:

> This is a minute proportion of all firms, and he provides no evidence that they employed a significant proportion of the labour force. Furthermore, it is clear that the application of the system rarely involved major job redesign, particularly for key groups such as craftsmen, and was frequently manipulated at shop floor level.
>
> (Batstone *et al.* 1987: 9)

Thus, as far as Britain is concerned at least, it would seem the idea that the traditional approach by management to the organisation and control of work can be characterised exclusively in terms of Taylorist practices needs to be treated with some caution.

Caution is also needed in interpreting more recent management approaches to the introduction and operation of new technology in terms of a 'Taylorist' or even 'neo-Taylorist' strategy to increase managerial control by deskilling work. For example, Wilkinson (1983; 1985) talks of managers in one of his case-study firms 'openly adopting a deskilling strategy' in their choice of equipment with the intention of increasing management control. In another he describes a debate within line-management and engineering functions over whether new technology should be used to retain or remove skill from the shop floor (1983: 89–90). This and other evidence leads him to suggest that managers often feel that 'current custom and practice represents a lack of control on their own part and try to remedy this by taking advantage of the opportunities lent by the new technology' (1985: 25). Similarly, a number of other commentators have suggested that despite there being alternative choices, new technology is generally seen by managers as a means of wresting control away from the shopfloor along what are in essence Taylorist lines (see for example, Child 1984: 252–7; Gill 1985; Baldry and Connolly 1986; Smith and Wield 1987).

We would argue that such observations may run the risk of misinterpreting the nature of managerial intentions when they seek to use new technology to increase control. This can be illustrated by looking at Buchanan and Boddy's conclusion, drawn from their Scottish case studies, that middle and junior line-managers were preoccupied with using new technology to satisfy what they term 'control objectives'. These objectives were a 'persuasive theme in discussions with middle and lower managers'. They were expressed as a 'desire for increased predictability, consistency, orderliness and reliability' in work operations and for a reduction in uncertainty in management control. Management expectations were that this could be achieved by increased reliance on computer controls and reduced dependence on what was regarded as the 'undesirable' control exercised by humans (1983: 244). If one assumes that Taylorism is an extensive feature of British management practice, then it is a short step to interpret the pursuit of 'control objectives' as a direct expression of 'neo-Taylorist' ideas. However, if one questions such an assumption then the nature of the intentions behind the pursuit of control objectives requires further investigation and explanation.

The nub of our dissatisfaction with Buchanan and Boddy's use of the term 'control objectives' is that it fails to distinguish between what might be termed 'labour control objectives' – aimed specifically at improving human performance and productivity – and 'operational control objectives' – aimed at the overall improvement of the performance of the production process itself (McLoughlin 1983). Whilst the pursuit of 'operational control objectives' may involve attempts to reduce the need for skilled or informed human intervention, managerial intentions cannot be reduced exclusively to a desire to deskill labour. In fact, other concerns may be more significant and in certain circumstances require a 're-skilling' or 'up-skilling' of labour.

This is illustrated particularly well in four recently published case studies by the late Eric Batstone and his colleagues[4]. Their findings suggest clearly that labour control was not a decisive factor in the decision to introduce new technology. The reasons for its introduction did, nevertheless, have implications for the process of labour regulation in so far as new technology was seen as a means for increasing managerial 'control'. The key point is that management attempts to increase its control were not driven by a concern to deskill labour *per se*, but rather a desire to improve the performance of the production process. For example, in the case of a chemical plant the key factor in management's attempts to reduce its dependence on human labour was their concern with quality, rather than a fear of any reliance upon workers directly. Similarly, in an insurance company the fact that new technology enabled the substitution of labour and reduced worker autonomy was not the sole criterion guiding managerial decisions but followed from the more central concern for reduction of administrative costs' (see Balstone *et al.* 1987).

We would offer similar interpretations of the nature and significance of control objectives in the cases of technological change studied by ourselves and discussed in the previous section (see McLoughlin 1983; Clark *et al.* 1988). For example, in the British Rail case the control objectives behind the introduction of the new technology were concerned primarily with increasing management control over physical resources – wagons and locomotives – rather than human resources. Moreover, control objectives were linked strongly to operating objectives aimed at improving the efficiency of freight operations and strategic objectives aimed at arresting the decline of the freight business. Similarly, in the case of British Telecom's exchange modernisation programme, although labour control objectives in the form of improved labour productivity and staff reductions were intended, these concerns were derived from, and were secondary to, operating objectives aimed at reducing costs and strategic objectives aimed at improving the range and quality of telecommunications services that could be offered to satisfy increasing customer demand.

The implications of the pursuit of control objectives by managers will be returned to in subsequent chapters. For the present it is sufficient to note that the general lack of a consideration at a strategic level of how work should be organised and controlled (what we have termed the 'control issues' highlighted by the introduction of new technology – see Chapter 2) can result in a further 'attenuation' of corporate objectives. In particular, where lower-level managers have no guidelines to the contrary, the decisions – or nondecisions – they make in relation to such things as job content, work organisation and supervision may in some circumstances frustrate the achievement of the overall objectives behind the introduction of new technology.

Conclusion

The empirical evidence reviewed in this chapter suggests that policies concerned with the regulation of labour do not flow unproblematically from overall business strategies. Such decisions involve political processes of strategic choice within organisations, which cannot be explained exclusively in terms of technical or commercial imperatives, or the logic of the historical development of forms of capitalist control over labour. It is apparent that managers pursue a diverse range of objectives when new technology is introduced. These reflect hierarchical and functional divisions and cannot be expressed in terms of a unitary strategy. Moreover, available evidence suggests that the introduction of new technology has not been accompanied by significant innovations in policies for the regulation of labour and that, in general, personnel specialists play a marginal and reactive role. In consequence, decisions to introduce new technology may leave considerable room for manoeuvre – or 'design space' – for lower-level managers at subsequent stages of change. 'Sub-strategies' developed by such managers in implementing and operating new technology can therefore have a critical bearing on the outcomes of change.

The cases of technological change in British Rail, British Telecom and the independent television company illustrate different approaches to implementation that can emerge and the ways in which these may facilitate or frustrate the achievement of senior managers' objectives. It was also noted that middle and junior line-managers are often preoccupied with the pursuit of 'control objectives' when new technology is introduced, but that these cannot be reduced to a concern to deskill labour along Taylorist lines as many labour process writers argue. The implications of these findings for job content and work organisation will be explored in Chapter 6, and for the supervision and control of work in Chapter 7. First, however, it is necessary to review the role of trade unions in the process of technological change and to explore in more detail the possible influence of new technology itself on changes in work tasks and skills.

Notes

1 Wilkinson's case studies were concerned with the introduction of microelectronics-based controls in a plating company, an optical company, a rubber moulding company and a machine tool manufacturer. These studies will be referred to at a number of points in the book.
2 For further explanation of notes in tables from the WIRS see Daniel 1987: xv–xvi.
3 The Bedaux system combined Taylorist principles with elements of industrial psychology and the fatigue studies conducted during the First World War aimed at improving industrial productivity (see Littler 1982; Rose 1978).
4 Batstone *et al.* studied four cases in all. These were drawn from brewing (automated process controls), small-batch engineering (CNC machine tools), chemicals (automated process–sequence controls) and the insurance sector of finance (on-line processing).

CHAPTER 4

Trade unions and technological change

Having considered management and management strategies, in this chapter we now turn to the question of the role of trade unions in technological change at work. The main focus will be on the union response to the new computing and information technologies. Again, trade union influence will be examined with reference to the various stages in the process of technological change. The new technology raises two kinds of bargaining issue for unions: procedural questions over when and how unions can influence the introduction of new technology, and substantive questions over which issues can and should be the subject of union concern and influence. Traditionally, decisions over whether new technology should be introduced, how equipment should be designed or which type chosen, have been left to management. Trade unions have therefore not sought to influence decision-making during the initial stages of change. Rather, they have been concerned to bargain over issues such as job security, pay and grading in order to negotiate a 'price for change' once these decisions have been taken and to ensure that those affected by change are protected and the benefits are widely distributed (see Mortimer 1971: 5).

However, at the end of the 1970s, a number of commentators, and indeed unions themselves, questioned the validity of this approach in the face of new microelectronics-based technologies. Not only were traditional issues such as pay and job security highlighted by computing and information technologies, but also other employment issues such as the potentially disproportionate negative effect on female employment (see for example, Arnold *et al.* 1982; Huws 1982; West 1982; TUC 1984; Werneke 1985; Greve 1986) and control and strategic issues concerning such things as the content of jobs and the manner in which unions could influence corporate management decisions (see for example Bamber 1980; Manwaring 1981; Wilkinson 1983; Winterton and Winterton 1985).

This chapter begins by discussing the view that trade unions have acted as a barrier to technological change in Britain. The policies developed during the late 1970s by the TUC in response to the wider diffusion of microelectronics are then examined, and the reasons for the failure of this initiative – based around the idea of *new technology agreements* (NTAs) – are outlined. Evidence on the effectiveness of trade union organisation and existing collective bargaining frameworks in negotiating technological change at work is then reviewed. As will be seen, in cases where change has been subject to negotiation, this has been limited both in terms of the stages of change at which bargaining has taken place and the range of substantive issues involved. Moreover, union responses have been hindered in some circumstances by a failure to come to terms with the erosion of occupational boundaries and skill demarcations that can accompany technological change. An attempt is then made to put the findings on the role of unions into a wider context by comparison with the process of technological change in non-union environments. Finally, trade union effectiveness is assessed by exploring evidence on the outcomes of change over the traditional bargaining issues of jobs and pay.

Trade unions: a barrier to innovation?

According to one recent commentary on the decline of the British economy since the Second World War, the trade unions have been 'possibly the strongest single factor militating against technological innovation and high productivity' (Barnett 1986; quoted from Daniel 1987: 182). Indeed, ever since the nocturnal machine-smashing exploits of the Luddites in the early nineteenth century, the image of unions as interested only in 'wrecking' employers' attempts to adopt new technology has persisted in popular, journalistic, political and many academic imaginations. Popular views have been bolstered by the reporting of highly visible and newsworthy disputes such as those in the national newspaper industry. Among politicians and policy-makers, too, there has been a longstanding belief that trade union activity, for example in the maintenance of restrictive practices, is a principal factor in explaining Britain's slow rate of technological innovation. Finally, these views have been given intellectual credence by some economists and innovation theorists who tend to see the role of trade unions in technological change as generally obstructive (see Willman 1986: 2; Wilkinson 1983: 9).

However, the view that trade unions in the post-war period have been a major barrier to technological change in general, and to the adoption of new computing and information technology in particular, can be questioned in a number of ways. First, throughout the post-war period, and most noticeably in recent years, the TUC and most member trade unions have given consistent support to the view that rapid technological change is an economic necessity. Second, there is no widespread evidence of fundamental resistance to technological change, whether in the form of strike action or through the influence of immutable restrictive practices. Third, there is little evidence that managements are dissuaded from introducing new technology by trade union attitudes. Finally, there appears to be a high level of support for the introduction of new technology among employees themselves.

An examination of TUC policy since the Second World War reveals consistent support for rapid technological innovation (see Willman 1986: 8–22). For example, in 1955 the TUC published a 'Memorandum on Automation' which was subsequently adopted by Congress and, with successive revisions, was the basis of TUC policy until 1979. This policy encouraged a 'willing acceptance' of technological change, giving support for innovation based on the view that any threat to jobs implied by automation could be countered by local negotiations designed to secure safeguards on job security. Subsequent revisions to TUC policy stressed the importance of employer and in particular Government policy in ensuring that rapid technological change was accompanied by the maintenance of full employment. Widespread public concern over the implications of new microelectronics-based technologies in the late 1970s prompted a major review of TUC policy in 1979. This involved a specially constituted Employment and Technology Committee, a special conference on technology, and the adoption of a major report, *Employment and Technology*.

These activities, according to Willman, constituted 'the most extensive discussion of technological change within the union movement in the post-war period' (1986: 12). The outcome of this policy debate reaffirmed the TUC's positive commitment to technological innovation with the proviso that a Government-sponsored industrial strategy, seeking to combine economic growth with full employment, was essential if the threat to employment posed by the new technology was to be countered. Whilst the existence of positive TUC support for technological change is not necessarily proof of union support for change in the workplace, the consistency of union leadership commitment to innovation since 1945 is noteworthy, especially given fluctuations in other policies and views during the period.

Statistics on the incidence of strike activity relating to the introduction of new technology since 1970 reveal that technology-related disputes, rather than being a common occurrence throughout the economy, have been concentrated in three industries – docks, printing, and motor vehicles – which accounted for the majority of days lost. Most other industries over the past decade appear to have been free from disputes over technological change (see Willman 1986: 47–50; Batstone and Gourlay 1986). However, there have been instances, for example in the Civil Service, where the existence of centralised computer installations – in particular in the provision of tax and social security services – have had a significant bearing on the conduct of disputes, providing unions with strategic points at which to apply pressure on the employer (see Beardwell 1987).

Nevertheless, there is little evidence to support the view that employers choose not to introduce new technology because of a fear of strike action. A CBI data bank survey, for example, revealed that firms which introduce new technology are no more prone to industrial disputes than firms that do not (cited by Dodgson and Martin 1987: 11). Similarly, as Willman points out, whilst there is well-documented evidence that restrictive practices exist in a number of industries, there is little data to suggest that these necessarily give rise to trade union resistance to technological change or that trade unions alone are responsible for the maintenance of restrictive practices. Even if it is accepted that the principal influence of restrictive practices is through 'featherbedding', i.e. where trade unions are able to negotiate the introduction of new technology on the basis that old and unsuited working practices are maintained – as occurred in national newspapers during the 1970s for example (see Martin 1981) – it is difficult to attribute such resistance that may occur exclusively to trade unions. In such circumstances the maintenance of restrictive practices has to be explained at least in part by managerial willingness to sanction or to overlook their existence in the first place (see Willman 1986: 54–9).

In addition, survey evidence on managerial perceptions reveals little concern over trade unions as a major obstacle to change[1]. The Workplace Industrial Relations Survey, for example, found that manufacturing establishments which recognised manual trade unions were more likely to have introduced advanced technology than their non-unionised equivalents (Daniel 1987: 34). Managers also reported that manual workers directly affected by the introduction of advanced technology were 'in favour' of change in seventy-five per cent of cases and 'strongly in favour' in fifty per cent of cases. Where resistance was reported managers said this was slight, and in only two per cent of cases did management regard workers' reactions as 'strongly resistant' (1987: 184). More significantly, managers judged the degree of support of shop stewards for the introduction of new technology as similar to that of the workforce in general and full-time trade union officials were perceived to be even more in support of change (1987: 187).

A similar picture is revealed when looking at survey evidence on the attitudes of union members and officials. An opinion poll conducted for the Technical Change Centre by MORI revealed that only a tiny proportion of trade union members surveyed saw technological change as a threat to be resisted, while twenty-one per cent thought new technology was beneficial and to be accepted willingly (cited by Dodgson and Martin 1987: 11). The WIRS found that manual shop stewards reported their initial reactions to be 'in favour' of change in over fifty per cent of cases, and subsequent reactions to be 'in favour' in over seventy-five per cent of cases. The stewards' perceptions of workers' views showed a similar, though less emphatic pattern of support for advanced technological change (1987: 194–5).

It would be wrong to interpret the above evidence as suggesting that trade unions, far from opposing change in the best traditions of 'Luddism' popularly attributed to them, always and without reservation wholly endorse management objectives. Clearly such a view is as absurd as that which stereotypes unions as anti-new technology. Paul Willman has recently advanced a complex and illuminating analysis of the sources of trade union resistance to technological change. His account draws an important distinction between two kinds of collective bargaining relationship. The first is what he terms 'sequential spot contracting' where the terms and conditions of the effort bargaining are constantly negotiated and re-negotiated at workplace level. The second is referred to as 'contingent claims contracting' where management and unions negotiate formal agreements to cover the terms and conditions of employment over specified periods of time (1986: 85–6). What distinguishes these two types of collective bargaining relationship is that the latter is more receptive to rapid technological change than the former, since spot-contracting offers continual opportunities to work groups to engage in sectional bargaining over the content and pace of change.

Significantly, spot-contracting has been the dominant form of collective bargaining in many of the British industries associated with union resistance to technological change – docks, printing, and the British-owned sector of the motor vehicles industry. Willman suggests that the unsuitability of spot-contracting as a basis for negotiating rapid technological change is the reason for union resistance in these industries, where managers wish to use new technology to cut costs to combat volatile product market conditions. It is also the reason why the introduction of new technology in these industries has been accompanied by attempts to move from collective bargaining based on spot-contracting to collective bargaining based on contingent claims contracts. Thus, as Willman notes, any explanation of trade union resistance to technological change needs to be based on an analysis of the links between product markets, management strategy, innovation policies and forms of collective bargaining (1986: 253). Clearly, this is a far cry from the 'Luddite' stereotypes of many popular and journalistic conceptions.

Trade union policy: the rise and fall of new technology agreements

As indicated above, while maintaining a commitment to technological innovation there was growing concern within the trade union movement during the late 1970s with the issues that were raised by new computing and information technologies. The *Employment and Technology* report adopted by the TUC Congress in 1979 recognised that, if the negative implications of the new technology were to be avoided, then unions would have to seek more effective ways of negotiating its introduction. This involved a renewed emphasis on bargaining over change around the idea of new technology agreements (NTAs) and the encouragement of government policies to support this process through existing tripartite bodies such as the National Economic Development Council (NEDC).

The idea of NTAs had originally been pioneered by trade unions in Scandinavia during the early 1970s (see Benson and Lloyd 1983: 155–64; Gill 1985: 141–60). Their adoption in Britain marked an attempt by the TUC to develop new approaches which increased union control over technological change as part of a general extension of industrial democracy. As David Lea, chairman of a special employment and technology committee established by the TUC and erstwhile member of the Bullock Committee on Industrial Democracy (1975–7), put it in 1980:

> The trade union response to technological development is clearly linked to the area of industrial democracy. We are seeking to both widen the agenda on which bargaining takes place and to exert influence at a much earlier stage in decision-making.

> (Quoted from Benson and Lloyd 1983: 166)

In other words, the TUC's response to new technology emphasised the need to move beyond bargaining over the employment issues raised by technological change to consider in addition the control and strategic issues that were also involved.

In seeking to put this response into practice, the TUC report recommended a checklist of ten points which trade union negotiators should seek to include in NTAs (see Figure 4.1). Four points referred to the procedural content of agreements. They emphasised that the procedures for change should be based on collective bargaining whose progress should be monitored by joint bodies; consultation with trade unions should begin prior to the decision to introduce new technology; information on management plans should be made available to the unions; and until such time as agreement was reached, the status quo should be observed. The other points concerned the substantive aspects of change. In particular, negotiators were advised to secure agreements on job security, the expansion of output, earnings levels, management commitment to retraining, a reduction of working hours, equipment design, and health and safety.

Partly in the wake of this initiative a number of unions, especially those representing white-collar workers, rapidly developed their own policy responses. Figure 4.2 provides a summary of the range of individual union policies (for further details see Manwaring 1981; Markey 1982; Davies 1986). White-collar unions such as APEX, BIFU, NALGO, and TASS tended to accept that new technology would lead to job losses, but accepted that resisting change would not help their members either. Their approach, therefore, was to attempt to ensure that the benefits obtained from new technology through improved productivity should outweigh the

Figure 4.1   TUC Checklist on New Technology Agreements

---

1. Change must be by agreement: consultation with trade unions should begin prior to the decision to purchase, and status quo provisions should operate until agreement is reached.
2. Machinery must be developed to cope with technical change which emphasizes the central importance of collective bargaining.
3. Information relevant to decision making should be made available to union representatives or nominees prior to any decision being taken.
4. There must be agreement both on employment and output levels within the company. Guarantees of job security, redeployment, and relocation agreements must be achieved. In addition, enterprises should be committed to an expansion of output after technical change.
5. Company retraining commitments must be stepped up, with priority for those affected by new technology. Earnings levels must be secured.
6. The working week should be reduced to 35 hours, systematic overtime should be eliminated, and shift patterns altered.
7. The benefits of new technology must be distributed. Innovation must occasion improvements in terms and conditions of service.
8. Negotiators should seek influence over the design of equipment, and in particular should seek to control work or performance measurement through the new technology.
9. Stringent health and safety standards must be observed.
10. Procedures for reviewing progress, and study teams on the new technology, should be established.

Source: Willman (1986)

obvious disadvantage of any job losses. In order to achieve this, the unions sought to extend their control over the implementation and price of change. The new technology agreement was seen as an essential means to this end. The most notable example here was the white-collar union APEX, whose national policy guidelines went further than most unions and indeed the TUC's, to include far-reaching recommendations on control issues such as the content of jobs and forms of work organisation (see Winterton and Winterton 1985; Francis 1986: 162).

However, the responses of other unions tended at one extreme towards positive support, and at the other to a concern to impose rigorous conditions before accepting change. For example, in British Telecom, unions such as the POEU (now

Figure 4.2   Scale of union agreement to change

Agreement		Disagreement
◀		▶
See advantages to members provided that certain minimum safeguards are met.	Realisation that new technology will lead to fewer jobs, but that resistance to productivity improvements would not help members. Extent to which the benefits outweigh the disadvantages depend on extent to which they can control the implementation and price of change.	Feel threatened by new technology and are only prepared to accept change subject to rigorous conditions.
(POE & EETPU)		*(NGA & ACTT)*
	(APEX; BIFU; NALGO; TASS)	

Source: Davies (1986)

the NCU) and the STE gave consistently strong general support for change (Clark *et al.*, 1988), while unions such as the EETPU saw positive benefits both for existing members and the possibility of union growth via recruitment and merger in the new 'information sectors' of the economy. The EETPU for example, saw printing work using the latest electronic technology as a legitimate job territory for its members whose skills, they claimed, were far more relevant than the mechanical craft skills of traditional print workers. It is not surprising, then, that the most cautious response to new technology should have come from craft unions, such as the NGA (see Gennard 1987), whose members' skills and jobs were most directly threatened by change.

However, notwithstanding the existence of these variations, in practice union bargaining over technological change – even where NTAs were a central plank of national policy – has tended to be based on existing procedures, and the effectiveness of NTAs where concluded has been limited. For example, Williams and Steward (1985) identified only 240 NTAs concluded between 1977 and 1983, suggesting that they had not been adopted widely during this period[2]. Moreover the signing of agreements appeared to have peaked in 1980 when sixty-five were concluded, followed by a marked decline in the rate of adoption (see Figure 4.3). Moreover, four white-collar unions, ASTMS, APEX, TASS and NALGO accounted for the majority of agreements (see Table 4.1).

More up-to-date and wide-ranging evidence has been provided by Batstone and Gourlay's survey of over 1000 shop stewards, which covered manual production workers, non-manual workers, and maintenance workers in both the public and private sectors. This indicated that over sixty per cent of non-manual workplaces, located mainly in the public sector, and twenty-two per cent of workplaces in 'production' were covered by new technology agreements (1986: 211–14). In general, the effectiveness of these agreements was regarded favourably by the stewards (see Table 4.2). However, Jary (1987) has suggested that these findings are rather optimistic given the rather 'loose' definition of a 'new technology agreement' adopted by the survey.

Perhaps the most significant indication of the effectiveness of NTAs, however defined, is the extent to which they have enabled the achievement of the TUC's original objectives of trade union influence at earlier stages in the process of change and a widening of the collective bargaining agenda to include control and strategic issues. Looked at in this way NTAs have fallen short of both objectives. Firstly, unions have had only limited success in securing a procedural right to influence management decision-making at a much earlier stage in the process of change. Secondly, they have not significantly widened the agenda on which collective bargaining takes place. For example, as Figure 4.4 suggests, only six per cent of NTAs surveyed by Williams and Steward stipulated union involvement from the initial planning stage of change, although forty per cent did involve an agreement that unions would be consulted before equipment was chosen.

Moreover, the degree of union involvement stipulated by these agreements fell a long way short of the TUC's recommendation that the status quo should be preserved until mutual agreement on the change had been reached. Just eleven per cent of agreements stipulated that new technology should be introduced only if there is mutual agreement (see Figure 4.5). However, over half did include clauses recognising that negotiations should take place prior to the installation of

Figure 4.3   Year of adoption of New Technology Agreements

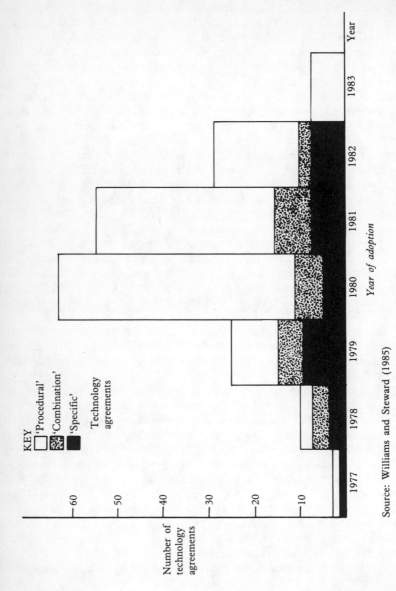

Source: Williams and Steward (1985)

Table 4.1 Trade union signatories to NTAs

	All technology agreements	Number of times union is signatory		
		Procedural agreements	Combination agreements	Specific agreements
APEX	66	50	12	4
ASTMS	63	56	5	2
AUEW/TASS	39	20	5	14
AUEW (Engineering Section)	6	6	0	0
ACTSS (Clerical Section of T&GWU)	15	7	6	2
TGWU	5	4	1	0
MATSA (Clerical Section of GMWU)	8	7	0	1
GMWU	4	4	0	0
EETPU	3	3	0	0
NUSMW (Staff Section)	2	2	0	0
Confederation of Shipbuilding and Engineering Unions (could include all of above)	2	2	0	0
BIFU	3	3	0	0
USDAW	8	7	1	0
NUJ	11	0	4	7
NGA	5	0	2	3
NATSOPA (Clerical Section)	4	0	0	4
ACTT	1	0	0	1
NATTKE	1	0	0	1
NALGO	53	47	1	5
NUPE	3	3	0	0
CPSA	1	0	1	0
SCPS	1	1	0	0
Council of Civil Service Unions (nine unions including two above)	2	0	1	0
NUR	1	1	0	0
TSSA	1	1	0	0
Other minor unions	3	3	0	0

Note: Because a given agreement may be signed by several unions the total of signings given above exceeds total number of agreements in survey

Source: Williams and Steward (1985)

Table 4.2 New Technology Agreements: numbers, coverage and usefulness

	Percentage with agreement	Coverage of agreement			Usefulness of agreement			
		Own union only	Comparable unions	Manual and non-manual unions	Very	Fairly	Not very	None
(a) Non-manual/public-sector group								
Finance	58	80	–	20	58	21	21	–
CPSA civil service	63	40	45	15	27	45	23	5
SCPS civil service	45	11	72	17	6	37	49	8
CPSA Telecom	97	93	–	7	26	26	39	10
POEU Telecom	57	95	2	3	20	51	18	11
(b) Production group								
Print	59	81	14	5	49	32	14	5
Chemicals	22	48	8	44	29	51	20	–
Food and drink	28	45	28	28	28	62	10	–
Engineering	15	33	54	13	33	40	14	13
(c) Maintenance group								
Chemicals	8	(50)	(50)	–	25	25	25	25
Food and drink	22	(83)	(8)	(8)	21	50	21	8
Engineering	13	(50)	(50)	–	19	19	31	31
Electrical engineering	21	14	43	43	–	84	16	–

Source: Batstone and Gourlay (1986)

Figure 4.4 Stage of union involvement stipulated in New Technology Agreements

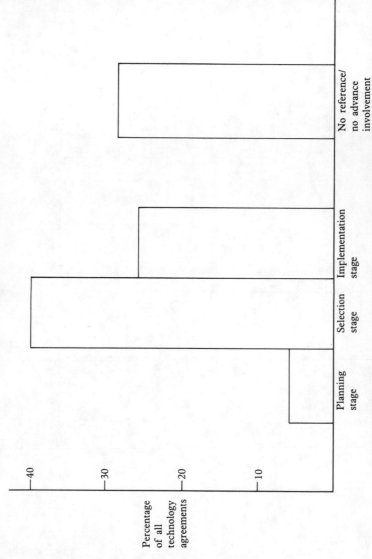

Stage at which unions are first involved in discussing technological change under technology agreements, distinguishing; Initial planning of technological change;
Selection of equipment/system;
Implementation of change.

Source: Williams and Steward (1985)

Figure 4.5   Degree of union involvement stipulated in New Technology Agreements

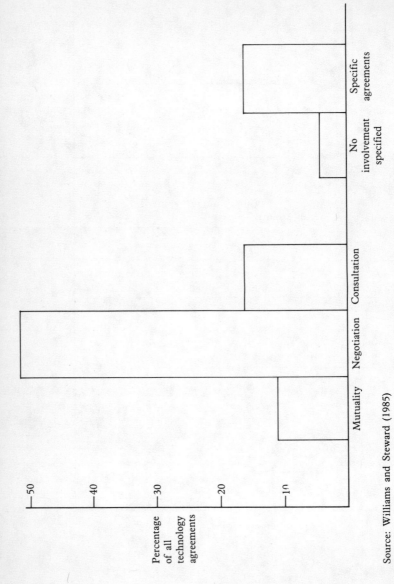

Source: Williams and Steward (1985)

equipment and changes in working methods. Finally, the vast majority of NTAs surveyed included substantive clauses on traditional employment issues such as job security and improved terms and conditions of employment, but only a minority included clauses on control issues such as skills and work organisation.

By the mid 1980s the new technology agreement initiative had lost its impetus, at both a national policy level and in the negotiating practice of individual unions. There are several reasons for this (see Gill 1985; Francis 1986; Batstone and Gourlay 1986; Dodgson and Martin 1987; Jary 1987). First, the policy had been conceived during a Labour administration in a framework of tripartism and national economic planning. However, it had to be implemented in the somewhat less convivial economic and political environment of increasing unemployment and a Conservative administration hostile to the trade unions. Second, employer attitudes influenced by government policy and increasingly competitive product markets have tended to encourage management resistance to the further extension of collective bargaining, if not promote attempts to reassert the managerial prerogative. (These two factors also encouraged the CBI to withdraw from a proposed joint statement with the TUC on technological change – see Benson and Lloyd 1983.) Third, the longstanding TUC commitment to technological change as inevitable and essential for economic growth tended to focus attention on employment issues and divert attention from issues such as the selection and design of equipment · or job content and work organisation. Fourth, the policy underestimated the organisational resources required for the official trade union bureaucracies, in particular in education and research, to service negotiators effectively. Finally, the strength of trade union organisation in the context of existing collective bargaining arrangements was not taken into account as a factor determining the effectiveness of union influence over technological change. The significance of this particular issue is returned to below.

So far then, it has been suggested that, contrary to popular stereotypes and the role conventionally attributed to them by some economists and innovation theorists, trade union policies and attitudes have in general been positive towards technological change. This fact has been reflected in the most recent policy initiatives of the TUC and its member unions towards the new computing and information technologies. These have shown a positive acceptance of the need for change, but have sought to minimise the negative effects on union members, in particular through the device of the new technology agreement. However, despite their rapid rise to prominence in trade union policy-making, the evidence reviewed suggests that new technology agreements have had little impact in practice.

The overall numbers of new agreements concluded appears to have been falling since a peak in 1983, while the content of those concluded has also fallen a long way short of the objectives of extending union influence over management decision-making and widening the agenda of bargaining originally set out by the TUC. The reasons for the fall of this policy in part reflect the changing economic and political circumstances since its inception. However, internal organisational factors have also been important; indeed, these factors have been crucially important in determining the degree of influence unions have been able to exert over technological change during the 1980s.

Collective bargaining and new technology: the negotiation of change?

This section reviews evidence on how far trade unions have been able to consult and negotiate over the introduction of new technology in the vast majority of cases where NTAs have not been agreed. In other words, how far have existing collective bargaining frameworks proved an adequate basis for the negotiation of change? First, evidence from surveys of managers and shop stewards will be reviewed, and then detailed examples from case studies will be outlined. Finally, the implications for trade unions of wider changes in occupational boundaries and skill demarcations will be considered.

Survey evidence on the negotiation of change

In the absence of the widespread adoption of NTAs, most unions, in particular those representing manual workers, have in practice sought to negotiate over technological change through existing collective bargaining arrangements. Batstone and Gourlay have argued that where existing trade union organisation is strong, then they can be expected to enjoy a wide range of influence over bargaining issues before new technology is introduced. This will be reflected in a union's capacity to exert control over the process and outcomes of technological change. The major factor determining the impact of new technology on workplace trade unionism, therefore, is existing trade union strength rather than changes in market conditions, management strategy, or the inherent features of the technology itself (1986: 38).

Evidence to support this view is provided by Batstone and Gourlay in their survey, already referred to above, of over 1000 shop stewards (1986: 185–211). They found that the extent of collective bargaining over technological change, in terms both of bargaining levels and the range of issues negotiated, was clearly correlated with the existing pattern of bargaining in particular sectors. This in turn was influenced by the strength of trade union organisation, expressed in terms of the 'sophistication' (for example, union density, numbers of shop stewards, existence of steward hierarchies) of workplace organisation and the degree of 'external integration' between workplace organisation and the official structure of the union.

Where union sophistication and external integration were both high, then the number of levels of bargaining and the range of issues negotiated in relation to technological change were also high. Where bargaining occurred on a number of levels a union could in principle influence both the formulation of management strategy in the initial stages of change by bargaining at company or national level and also influence the implementation of new technology at workplace level during subsequent stages of change. On this basis Batstone and Gourlay concluded that multi-level bargaining is the most effective means for trade unions in seeking to influence technological change since it allows negotiation at different stages and at different levels during the process of change.

However, Batstone and Gourlay's evidence indicates that, in practice, the incidence of such multi-level bargaining in manufacturing is very limited. For example, only twenty-five per cent of shop stewards surveyed in this sector claimed the union engaged in multi-level bargaining over technological change at two levels, and only ten per cent at three levels. Multi-level bargaining was more likely to be reported by stewards in the non-manual and public sector groups surveyed. Here over fifty per cent of respondents said they bargained at more than one level (Batstone and Gourlay 1986: 191). Nevertheless, it also appears that in most cases, irrespective of the number of levels at which it took place, bargaining was largely restricted to employment issues such as manning, pay, training and health and safety, and did not cover control issues or strategic issues such as investment or equipment (see Table 4.3).

Table 4.3　Areas of negotiation over technological change

	Invest- ment	Equip- ment	Manning	Pay	Training	Health
(a)　Non-manual/public-sector group						
Finance	–	30	70	70	27	84
CPSA civil service	5	51	88	67	67	90
SCPS civil service	7	23	87	54	57	89
CPSA Telecom	17	77	100	80	71	97
POEU Telecom	40	52	75	82	34	59
(b)　Production group						
Print	16	31	80	75	58	63
Chemicals	17	21	83	64	41	79
Food and drink	23	33	82	92	54	77
Engineering	16	16	75	68	52	57
(c)　Maintenance group						
Chemicals	12	5	51	45	41	60
Food and drink	10	8	57	44	53	51
Engineering	7	7	51	42	56	51
Electrical engineering	6	16	63	63	44	63

Source: Batstone and Gourlay (1986)

A broader picture of the extent of consultation and negotiation over technological change is provided by the Workplace Industrial Relations Survey (Daniel 1987: 112–50). Managers' responses revealed a high level of consultation with manual unions and employees, over ninety per cent reporting some form of consultation, and fifty-two per cent reporting consultation with shop stewards (see Table 4.4). However, change had been the subject of negotiation and agreement with manual unions in less than one out of ten plants. Even in large workplaces with a hundred per cent union membership, circumstances in which workplaces manual union organisation might be deemed to be at its strongest, change was the subject of a negotiated agreement in only twenty-seven per cent of cases. Significantly, management tended to engage in both consultation and negotiation only where pressured to do so by a strong union presence or adverse workforce reactions to change. Responses from manual shop stewards tended broadly to confirm this picture, suggesting a lower level of consultation but roughly similar levels of negotiation. In the case of office workers there appeared to be a higher propensity

Table 4.4 Comparison of managers' and manual stewards' accounts of consultation affecting manual workers

Column percentages[a]

| | Union recognised | | Establishments where both were interviewed about advanced technical change | |
	All managers reporting advanced technical change	*All stewards reporting advanced technical change*	*Managers*	*Stewards*
One or more of listed forms of consultation	91	84	95	81
No consultations	8	14	4	17
Not stated	1	2	1	2
Formal – union channels[b]				
Discussed with shop stewards	52	41	59	46
Discussed with full-time officers	21	17	30	18
Formal – non-union channels				
Discussed in established JCC	20	24	29	20
Discussed in specially constituted committee	13	17	21	20
Informal				
Discussion with individual workers	62	49	63	50
Meetings with groups of workers	42	39	49	42
Base: see footnote[c]				
Unweighted	*405*	*288*	*205*	*205*
Weighted	*160*	*101*	*72*	*72*

[a] See note C
[b] See note F
[c] In the first two columns we compare the accounts of all managers who reported advanced technical change in places where manual unions were recognised with the reports of all the manual shop stewards who reported such change. In the second two columns we confine the analysis to cases where managers and stewards both reported advanced technical change. The first contrast shows the picture that would have been revealed had we depended solely upon the accounts of managers, compared with the results had our only source of information been the reports of stewards. The second contrast highlights differences between management and union perspectives, perceptions or recollections when talking about the same events in the same workplace.

Source: Daniel (1987)

for management to consult through non-union channels, although the extent of negotiation over change was similar to manual workers. However, management and stewards' accounts tended to be at some variance, with managers reporting higher levels of consultation and shop stewards' higher levels of negotiation over change (see Table 4.5).

Table 4.5 Comparison of managers' and non-manual stewards' accounts of consultation in the office

Column percentages[a]

| | Union recognised | | Establishments where both were interviewed about an advanced technical change | |
	All managers reporting advanced technical change	All stewards reporting advanced technical change	Managers	Stewards
One or more of listed forms of consultation	94	79	94	83
No consultations	4	19	2	16
Not stated	2	2	4	1
Formal – union channels[b]				
Discussed with shop stewards	35	34	49	36
Discussed with full-time officers	16	20	20	17
Formal – non-union channels				
Discussed in established JCC	17	30	23	32
Discussed in specially constituted committee	16	16	18	18
Informal				
Discussions with individual workers	73	50	71	50
Meetings with groups of workers	46	34	53	33
Base: as column heads[c]				
Unweighted	*747*	*520*	*437*	*437*
Weighted	*286*	*201*	*138*	*138*

[a] See note C
[b] See note F
[c] See note to Table vi.7

Source: Daniel (1987)

Case-study evidence on the negotiation of change

Case-study evidence presented by Jary (1987) suggests some support for Batstone and Gourlay's model in so far as existing collective bargaining arrangements, as shaped by union sophistication and integration, tended to be the vital factor in the extent of union influence. For example, in one case involving the introduction of robot spot welders on to Ford's new 'Sierra' production line at Dagenham, it was found that the existing bargaining structure and collective agreements meant that formal negotiation took place only at company level between management and national union officials. In addition, a national agreement had been signed some years earlier committing the manual unions within Ford to full co-operation with the introduction of new technology.

Thus, despite the existence of a relatively sophisticated joint shop steward organisation in the plant, they were prevented by existing collective agreements from negotiating during the implementation stage of change. Moreover, the low degree of integration with the wider union structure resulted in shop stewards having a lack of information on management plans, the capabilities of the technology, and the alternative bargaining options open to them. This meant that AUEW and EETPU stewards in particular were in a weak position to take the initiative, and in retrospect felt it could have been possible to reduce the impact of the 400 job losses in the areas of the production line in which robots were introduced.

A second case discussed by Jary, involving the introduction of a completely new production plant by a power cable company, revealed a contrasting situation. In this case union sophistication and integration between workplace and wider union structures was high and bargaining took place at several levels within the company. When it came to the negotiation of staffing levels and working practices for the new plant, the union (the TGWU) was able to secure a comprehensive and very favourable agreement with management before the plant was brought into operation.

Other case-study evidence confirms the general picture painted by survey evidence, which shows that in most cases the weakness of union organisation and bargaining structures tends to limit the extent of union influence over change. Moore and Levie (1985), for example, reported on union influence in GEC, British Leyland, Midland Bank, and the engineering company Alfred Herbert. In none of the cases was the decision to invest in or the choice of technology subject to negotiation. Collective bargaining took place over the effects of the technology on issues such as redundancies, job descriptions and changes in working practices, but not on the choice or design of technology.

The failure to negotiate could not be explained exclusively in terms of the hostile economic and political environment or more aggressive management attitudes. Rather, technological change highlighted existing weaknesses in trade union structure, organisation, activity and servicing far more sharply than other collective bargaining issues. Technological change appeared as yet another issue to be discussed on an already overloaded collective bargaining agenda, management decisions on technological change were taken at levels at which collective bargaining did not take place, and unions were not provided with timely information on management plans. In addition, inter-union rivalry led to divisions between unions at workplace level in an effort to defend jobs against other sections of the workforce, sometimes in the same union. Finally, weaknesses in links between shop stewards and other union members, and between stewards and full-time officials, were also exposed. Using existing collective bargaining arrangements as a basis for handling technological change was described by one trade unionist as 'like fitting a six foot corpse into a five foot coffin!' (1985: 514.)

Similar conclusions were reached by Wilkinson in his case studies in the West

Midlands. He found that issues of work organisation and skill, although critically important at workplace level, were not negotiated in the conventional sense. Official trade union concerns were with traditional employment issues such as payment, redundancy and safety. For example, in the case of a rubber moulding company introducing electronically controlled moulding machines, the machine operators' union (the AUEW) was concerned primarily to preserve its members' jobs and insisted that the new machines should be staffed by existing operators – a point conceded in negotiation by management. However, management's intention to allocate the new tasks of setting and adjusting the electronic controls to white-collar workers, effectively transferring skill from the operators to another area, went uncontested by the union (1983: 48–54). Wilkinson concluded pessimistically that, 'traditional methods of bargaining' were 'wholly inadequate for technical change' (1983: 99).

Divisions within the workforce: changes in occupational boundaries and skill demarcations

The new task and skill requirements generated by new computing and information technology, and the choices opened up in relation to job content, work organisation and supervision, are discussed in detail in Chapters 6 and 7. In general terms one potential major effect is to disrupt established occupational boundaries and skill demarcations (Winterton and Winterton 1985: 6) and in some instances to challenge traditional boundaries between male and female jobs. Evidence on the extent of changes in demarcation is provided by Batstone and Gourlay's shop steward survey. This revealed that a great deal of change had occurred, in particular in maintenance jobs, but also among production and some categories of clerical worker (1986: 221–3). Similarly, the WIRS found that manufacturing establishments using advanced technology were far more likely to have flexible working practices in both production and maintenance areas (Daniel 1987: 172–7).

The erosion of traditional job demarcations and management attempts to promote flexible working pose an additional problem for trade unions in attempting to negotiate over technological change. The removal of traditional boundaries between jobs and grades can spark off sectional divisions within the membership of a union and divisions between trade unions with competing interests. An example of this problem was noted in Chapter 3 in the introduction of ENG by the independent television company. This provided a clear illustration of the way the skill requirements of a new technology can threaten to provoke divisions within the workforce by eroding occupational boundaries (see Clark *et al.* 1984 for discussion). In this particular case, aided by a management prepared to adopt a highly participative approach to implementing change, the union concerned was able to devise a response which overcame the potential division between the two occupational groups – video and film technicians – over who should operate the new technology.

However, other case-study evidence suggests that in general trade unions have not been so successful in preventing damaging sectional divisions from undermining their response to technological change. For example, it has already been noted how inter-union rivalry weakened the trade union response in the firms studied by Moore and Levie. A major source of division in these cases was the attempt of craft unions representing skilled male manual workers to defend their job security and status by ensuring that no gains were made by unions representing semi-skilled workers. The energies exerted by the unions in fighting these sectional battles effectively meant that 'while management changed the rules and relocated the pitch, the unions were happy to continue fighting among themselves back in their own backyard' (1985: 520).

A clear illustration of the manner in which the effectiveness of trade union attempts to bargain over change can be undermined by sectional divisions is provided by Leach and Shutt (1985). They document the failure of trade union strategy in two firms in the cereals and snacks sector of the food and drink industry – Kelloggs and Smiths Crisps (subsequently taken over by Nabisco). In both cases the technology involved was a new computer–weigher which transformed product weighing and packing lines by increasing capacity while at the same time reducing the requirement for labour. One effect of the introduction of the technology was to further exacerbate overcapacity problems in the industry in the wake of falling consumer demand in the recession. Thus, large-scale plant closures were also a likely outcome.

In the case of Smiths Crisps, management pursued a strategy of introducing new technology, closing down old factories and reducing labour costs. The union response was led by Smiths National Joint Shop Stewards Committee, which planned a campaign against the closures organised by the main union, the TGWU. In addition, there was the possibility of truck drivers being called out on strike in order to paralyse distribution. The overall objective was to secure a new technology and job security agreement, while in the meantime refusing to accept new equipment without mutual agreement. In the event the strategy failed and the closures took place as planned without opposition.

The researchers suggest that 'the crunch' came for the following reasons. First, the unions had no overall strategy which covered all the company's plants in order to confront management's corporate plans, a problem exacerbated by management's unwillingness to provide information on anything other than a piecemeal basis. Second, the union organisation was ineffective, decisions taken nationally by the Combine Committee were overturned at branch level, and little information was provided to union members. Finally, eighty per cent of the Smiths Group workforce were women, many of whom worked part-time, while the majority of full-time officials were men with an over-representation from depots rather than the factories where closure was threatened.

A similar tale of division was found at Kelloggs. Here, according to Leach and Shutt, management successfully exploited divisions between craft workers and operatives, and male operators and female packers, to achieve the introduction of new technology. In this case the strong organisation of the craft workers represented by the AUEW enabled them to achieve a job security agreement. At the same time, the largest union USDAW was weakly organised, had a predominantly female membership, but mainly male activists. They were refused a job security agreement, and could do little to prevent management's plans for redundancies among female packers and the introduction of temporary employment contracts for a significant number of the remaining workforce.

To summarise: in the context of the failure of the new technology agreement initiative, trade unions have had to rely in most circumstances on existing collective bargaining arrangements as the basis for negotiating the introduction of new technology. The general picture to be gleaned from available evidence is that the success of unions in negotiating change is strongly related to the strength of existing trade union organisation. Apparently, where unions already enjoy influence over a wide range of collective bargaining issues, this influence is likely to continue if and when new technology is introduced. Where, on the other hand, trade union organisation is weak prior to technological change, this is reflected in their incapacity to negotiate effectively over new technology when it is introduced.

The most typical scenario is likely to have been one in which new technology has been introduced where trade union organisation has been weak. Moreover, even where trade union organisation has been strong, the range of issues covered has been limited to traditional concerns. Finally, as if these difficulties were not enough, weaknesses in union organisation have been further exposed by the tendency of technological change to disrupt established skill demarcations and occupational boundaries which can result in sectional divisions within and between unions. Does this mean, however, that the presence of trade unions is of little benefit to workers affected by technological change? The next two sections attempt to put the influence exerted by trade unions in context by comparison with the process and outcomes of technological change in non-union workplaces.

Technological change in the non-union firm

It has already been noted that the WIRS data revealed that non-union workplaces are less likely to have adopted microelectronics technology than their unionised counterparts. While this tends to undermine the simplistic argument that unionised firms are at a disadvantage to non-union firms when introducing new technology, it does leave open the question of whether the absence of unions makes the management task of introducing technology easier. In particular, it could be suggested that the lack of a union presence means that management is spared time-consuming exercises of consultation and negotiation and can therefore undertake the process of technological change far more rapidly by more or less imposing their decisions on the workforce. On the other hand it could be argued that the introduction of new technology, like any other major change at work, relies for its success on the consent and commitment of the workforce and that management will need to develop relatively sophisticated joint procedures to enable this to take place.

Data from the WIRS allows the construction of a broad picture of the way advanced technological change involving microelectronics-based technologies has been dealt with in non-unionised workplaces (Daniel 1987: 119–21; 137–8). In the case of manual workers the likelihood of there being no consultation with the workforce was far higher than in unionised workplaces (forty-four compared to eight per cent). Moreover, in the cases where some form of consultation took place this tended to be on an informal basis through discussions with individual workers or group meetings rather than formal channels such as joint consultative committees. In the case of non-unionised workers in offices the likelihood of there being no consultation with the workforce was again higher than in unionised workplaces (fifteen compared to four per cent). However, it was far more likely for management to consult non-unionised office workers than it was to consult non-unionised manual workers (eighty-three compared to fifty-four per cent).

Nevertheless, despite the higher level of consultation with non-unionised office workers, it was still the case that this tended to take place through informal channels rather than formally constituted joint procedures.

Unfortunately there is little case-study research which has examined the process of technological change in non-union environments which would allow comparison with unionised counterparts. However, the experiences of four firms studied by one of the present authors is instructive (see McLoughlin 1986; 1987; 1988). The firms concerned were all involved in the introduction of CAD equipment into their drawing offices. Two firms, a mechanical engineering company and a shipyard, were unionised, while the other two, a building firm and an engineering consultant, were non-union. The drawing office union TASS (at the time of writing about to merge with ASTMS), to which drawing office staff in the two unionised companies belonged, was one of the four white-collar unions which led the way in seeking to negotiate NTAs (see AUEW–TASS 1979; 1985).

The cases of the mechanical engineering firm and the shipyard tend to support the pessimistic view of the effectiveness of NTAs in practice. In fact, the TASS shop stewards in the mechanical engineering firm did not even seek to negotiate a new technology agreement with management despite this being national policy. Management therefore took the initiative and offered a no redundancy undertaking and retraining for all drawing-office staff (both provisions in TASS's national guidelines on negotiating technological change), but refused to offer extra payments for retraining or to allow any formal consultation over change through union channels.

At the shipyard a local new technology agreement was negotiated as an adjunct to existing national and company level collective bargaining arrangements in the industry. The substantive clauses of the agreement included commitments to no redundancies and no compulsory shiftworking, while procedural clauses included a requirement for consultation with the unions 'at the earliest stage'. In the case of CAD a joint working party, comprising drawing office managers and staff, was established to evaluate various CAD systems. In addition, during implementation a joint monitoring committee met on average once every six weeks. The strong position of TASS in the company also enabled the negotiation of a favourable pay and training package as a condition for accepting the new system. Compared to the mechanical engineering case the influence of TASS appeared far greater. However, this owed more to existing collective bargaining arrangements than to the new technology agreement, which was seen by management as of little consequence. These findings are supported by other studies of NTAs covering CAD (see Arnold and Senker 1982; Baldry and Connolly 1986).

Whilst findings such as these could be seen as supporting rather pessimistic conclusions about the extent to which TASS was effective in negotiating change, it is important to place the efforts of local union negotiators in the context of the experience of the workforce in the two non-union firms. The firms concerned were owner-managed and as a matter of principle did not recognise trade unions. In both cases decisions over the introduction of new technology at all stages in the process of change were taken by a small group of senior managers with the minimum of communication or consultation with their staff. At the consulting engineers the lack of information provided to staff led to unfounded rumours 'on the company grape vine' that redundancies would result from the introduction of the new equipment and this had a negative impact on staff morale. At the building firm the lack of consultation led to a dispute over new contracts of employment with the drawing office staff. In this case, management's approach to industrial relations was described candidly by one manager as 'a total cock-up!'

In fact, both sets of managements felt in retrospect that more could have been done to inform staff of their plans, although significantly neither was particularly keen to invest organisational resources in developing procedures to facilitate this. In the case of the building firm, for example, even the idea of a single half-hour briefing session was considered impractical because it would be too expensive! In both cases it is worth noting the outcomes of change. In both firms staff were selected for retraining at management discretion. At the consulting engineers no additional payments were made for retraining. At the building firm the 'carrot' of premiums (amounting to a fifty per cent pay rise) was used to impose a twenty-four hour, three-shift system – a pattern of working usually resisted in TASS organised drawing offices – on those selected for retraining.

The outcomes of negotiation: employment, pay and trade union organisation

The evidence presented so far suggests that trade union influence has been restricted in general to negotiating over a narrow range of employment issues in the later stages of the process of technological change. But what were the outcomes in such cases and how do they reflect the extent of influence on the part of the unions, be it weak or strong, at least in relation to traditional bargaining issues?

Reports from management and shop stewards surveyed by the WIRS on the impact of new technology on manual workers revealed that in over two-thirds of cases no change in employment levels occurred (Daniel 1987: 210–26). Where change had occurred, increases in employment levels were reported in about one-third of cases, and decreases in employment in the remaining two-thirds (see Table 4.6). There was no evidence that unionisation led to more favourable outcomes, and in workplaces where unions were not recognised, employers tended to increase rather than decrease staffing levels. In the case of offices, about thirty per cent of workplaces had experienced some change in employment levels due to the introduction of microelectronics technology. Here, numbers increased in around one-third of cases and decreased in about two-thirds (see Table 4.6). Where unions were recognised, there was more likely to be a reduction in staffing than in non-unionised offices. Batstone and Gourlay's survey paints a slightly different picture in that over twenty-five per cent of respondents reported job losses of ten per cent or more, although fifty per cent still said that technological change had either a negligible or no effect on employment in their firms (1986: 175).

The WIRS also allows an instructive comparison between new technology and other causes of job loss. Of the forty-three per cent of workplaces using new technology which reported a reduction in their manual workforce during the year prior to the survey, only nine per cent claimed this was due to the adoption of microelectronics. Reorganisation, lack of demand, and cash limits imposed by central government were all more frequently cited reasons for job loss. Similarly, of the forty-four per cent of workplaces using computers and word processors

Table 4.6   Impact of new technology on manning levels in manual sectors compared with offices

Column percentages

	All relevant establishments		Establishments where advanced technical change affected both categories of employee	
	Manual workers	*Office workers*	*Manual workers*	*Office workers*
Manning level in affected section(s)				
Increased	11	10	14	11
Decreased	19	18	29	26
No change	70	70	57	62
Not stated	*	2	*	1
Base:				
Unweighted	*458*	*977*	*326*	*326*
Weighted	*212*	*500*	*78*	*78*

Source: Daniel (1987)

reporting a reduction in office jobs, only six per cent claimed this was a result of the introduction of new technology. In these cases reorganisation, lack of demand, cash limits, and the need to reduce costs, were all more frequently cited reasons for workforce reduction.

When compared with workplaces where no new technology had been introduced, those adopting microelectronics were only slightly more likely to have experienced job losses. However, reports on general changes in the total number of people employed between 1979 and 1984, irrespective of whether new technology had been introduced or not, revealed that ninety-one per cent of workplaces had experienced change, fifty-three per cent a decrease and thirty-eight per cent an increase. Larger workplaces employing large numbers of manual workers were more likely to have experienced reductions than smaller workplaces, private manufacturing establishments tended to have experienced most decline in jobs, while private services had enjoyed a slight increase.

Significantly, there was a strong and consistent tendency for decline in the size of workforce to be associated with the recognition of trade union, and for increases to be associated with non-union status. Similar associations between job loss and the extent of trade union organisation were found by Batstone and Gourlay. Even more significantly, they found that job losses in some sectors were more likely to occur in areas where unions were well represented and well organised. They explored two possible reasons for this. First, a 'conspiracy theory', namely that managements were using the new technology to reduce trade union power and influence. However they found no evidence that managements were pursuing labour regulation policies of this type, a finding that is consistent with the research discussed in Chapter 3. An alternative possibility was simply that strong and active unions tended to be found in the workplaces most central to an organisation's activities, and it was precisely in these areas that management was likely to want to use new technology to improve competitiveness and efficiency (1986: 177).

Turning to pay, the WIRS (see Daniel 1987: 242–50) revealed that in the majority of cases there was no immediate change in the earnings of manual and non-manual workers directly affected by new technology. This finding is also broadly supported by Batstone and Gourlay's survey which also found that in most cases technological change had no effect on pay and grading. According to the WIRS survey, in the case of unionised manual workers there was no change in earnings in seventy-two per cent of cases, and in the workplaces where change did occur this tended to be in the form of an increase (twenty-four per cent of cases) rather than decrease (three per cent of cases). In the minority of workplaces where earnings did increase, it was often the result of negotiated agreements reached as the basis for accepting technological change. When compared with non-union environments the pattern was more-or-less similar, suggesting that unionised workers have not fared significantly better (see Table 4.7).

Table 4.7 Impact of new technology on manual workers' earnings in relation to union recognition

Column percentages

	All establishments		Medium size establishments with 50–199 manual workers	
	Manual union recognised	*Union not recognised*	*Manual union recognised*	*Union not recognised*
Implications of change for earnings				
Increased	24	24	24	(50)[a]
Decreased	3	–	3	–
No change	72	76	72	(50)
Not stated	1	–	1	–
Medium/source of any increase[b]				
Regrading	9	11	9	(12)
Bonus payments (PBR)	10	5	12	(13)
Change agreements (increased rates)	9	5	12	(14)
More overtime	5	3	10	(9)
Other answer	3	7	2	(19)

Base: establishments with 25 or more manual workers and experiencing advanced technical change

Unweighted	*405*	*53*	*100*	*26*
Weighted	*160*	*52*	*64*	*18*

[a] See note B
[b] See note F

Source: Daniel (1987)

In the case of office workers a less favourable situation for unionised employees was reported (see Table 4.8). Here no change occurred in seventy-seven per cent of unionised workplaces, increases in twenty-one per cent of cases, and decreases in no cases. However, in non-union workplaces, no change was reported in seventy per cent of cases, increases in twenty-seven per cent, and a decrease in no cases. The pattern was markedly different in smaller establishments employing under fifty non-manual workers and in larger establishments employing over 150 non-manual workers. Here non-manual workers tended to fare better than their unionised counterparts. In smaller workplaces non-union employees enjoyed increases in thirty-one per cent of cases compared with seventeen per cent for unionised office workers, and in larger workplaces increases occurred in twenty-nine per cent of non-union cases, compared with twenty-five per cent of unionised cases. The most common source of pay increase in all cases was the regrading of work, although negotiated increases were important in unionised workplaces. As the author of the survey observed, 'it certainly appeared that trade union representation did little to increase the chances that office workers had of increasing their earnings as a result of technological change' (Daniel 1987: 250).

Finally, evidence on the outcomes of technological change for trade union organisation can be considered. In the areas where new technology affected work directly, Batstone and Gourlay found little evidence of dramatic changes in union organisation. For example, three-quarters of shop stewards in the areas where new equipment was introduced reported that there had been no change in the union's role compared with areas which had not experienced technological change. More generally it was found that technological change had had no marked effect in most sectors of employment on a range of variables, including membership activism, unity between different unions, shop steward authority, and union control over work organisation. However, where change had occurred this tended to reflect the strengths and weaknesses of the existing trade union organisation. That is, 'the process of technical change tends to augment those characteristics of workplace union organisations that were apparent before technical change took place. Stronger organisations tend to become stronger, weaker organisations weaker' (1986: 262).

To summarise: this and the previous section have taken into account the non-union dimension in an attempt to assess the influence exerted by trade unions over technological change. The findings reviewed provide a mixed picture. On the one hand, unionised firms are more likely to introduce new technology than non-union firms, and it appears that in practice, even where union organisation is weak, the influence exerted over management is in excess of that which employees in non-union workplaces can hope to achieve. On the other hand, the outcomes of change in terms of the issues of job security and pay – the concerns traditionally most central to unions in collective bargaining – reveal that unions have not in many instances been able to offer demonstrable benefits over their non-unionised counterparts. If anything, non-unionism appeared to be associated with more jobs and higher pay. Whether this occurred because of the location of non-unionised workforces in high-growth sectors of the economy, the failings of union organisation, or management strategies designed to avoid unionisation, is a question which requires further investigation[3].

Table 4.8 Impact of new technology on earnings in relation to union recognition in offices

	All establishments		Small establishments with lower than 50 non-manual workers		Large establishments with 150 or more non-manual workers	
	Non-manual union recognised	Union not recognised	Non-manual union recognised	Union not recognised	Non-manual union recognised	Union not recognised
Implications of change for earnings						
Increased	21	27	17	31	25	29
Decreased	*	–	–	–	*	–
No change	77	70	79	64	74	71
Not stated	2	3	4	5	1	–
Medium/source of any increase[a]						
Regrading	16	20	14	20	19	25
Bonus payments (PBR)	1	–	–	–	–	–
Change agreements (increased rates)	6	2	3	3	8	*
More overtime	*	4	–	3	*	5
Other answer	3	6	3	7	3	3

Base: establishments with 25 or more non-manual workers introducing computers or word processors

Unweighted	747	230	55	74	482	76
Weighted	286	213	83	119	84	23

[a] See note F

Source: Daniel (1987)

Conclusion

New technology agreements were designed to both broaden the range of issues and extend the influence of trade unions into earlier and higher levels of management decision-making. In practice NTAs have not been widely adopted, and where they have their form has fallen some way short of the original TUC objectives. At best, trade unions appear to have been able to exert most influence where the existing collective bargaining framework enables them to exert a strong influence over other issues. This is most likely where multi-level bargaining already exists, enabling negotiation to take place at company level over strategic issues during the initial stages of change, as well as at the workplace during the implementation stage. A key factor here appears to be the degree of sophistication of union workplace organisation and the extent to which it is integrated with the wider official union structure. However, it appears that even in these circumstances the scope of bargaining is still restricted to traditional employment issues such as job security, and pay and grading. At worst, trade union influence has been limited by weaknesses in union organisation and existing collective bargaining arrangements.

A comparison of the process and outcomes of change between unionised and non-unionised workplaces provides considerable food for thought for trade unions – in particular where, on the most fundamental issues of job security and pay, unionised workforces do not seem to have a demonstrable advantage over their non-unionised counterparts when it comes to negotiating technological change. Evidence such as this would appear to lend credence to the view that, especially where trade unions are already in a weak position, existing collective bargaining arrangements are an inadequate basis for the negotiation of change. On the other hand, there is little evidence to suggest that one outcome of technological change has been to weaken directly trade union organisation at the workplace.

Do these findings mean, however, that the workforce has little hope of influencing any of the control issues highlighted by the introduction of new computing and information technologies? As will be seen in Chapter 6, evidence suggests that in many cases it is individual workers and workgroups rather than unions which have exerted the most significant, and sometimes decisive, influence over issues such as job content, work organisation and the control of work.

Notes

1 As indicated in the Introduction, managers' perceptions of the barriers to innovation also tend to rate other factors such as the shortage of skilled personnel as far more significant than opposition from unions and workforce (see Figure A, Appendix).

2 Williams and Steward distinguish between three types of new technology agreement: 'specific' concerning the introduction of a particular system or item of equipment; 'procedural' which establish a framework for handling technological change in general; and 'combination' which combine both approaches. Of the 240 agreements surveyed, 39 were 'specific', 171 'procedural' and 30 'combination'.

3 In fact, these questions are under investigation as part of a research programme being carried out under the joint direction of one of the present authors, at the Kingston Business School, Kingston Polytechnic.

CHAPTER 5

New technology, work tasks and skills

In Chapter 2 it was noted that writers adopting both labour process and strategic choice approaches reject strongly the idea that technology is a primary explanatory variable determining the outcomes of technological change. In so doing, they also tend to play down, at an analytical level at least, any independent influence that technology may exert. This chapter begins by outlining a number of reasons why a refusal to take into account a detailed analysis of the independent influence of technology is seriously misguided.

In this context Woodward's insight that technology can have an independent influence on work and organisations is of considerable value. However, we believe that her approach to defining 'technology' is, as that of many other writers, inadequate for this task. The chapter therefore goes on to propose an alternative definition developed by ourselves and colleagues. As will be seen, one advantage of this approach is that it can be used to trace in detail the independent influences that particular technologies may exert. The remainder of the chapter explores the independent influence of new computing and information technologies on work tasks and skills, using detailed evidence from two case studies in which we have been involved. The suggestion to be made is that the capabilities and characteristics of the new technology may generate contradictory imperatives which both 'deskill' work tasks, in the sense of reducing the requirement for manual skills and dexterity, and 'upskill' work tasks in so far as new mental skills and abilities are needed if the new technologies are to be used effectively.

Beyond technological determinism

According to Braverman, in capitalist societies machines have two functions. The first is to increase the productivity of labour and the second is to deskill work in order to increase management control (see Chapter 2). Technology, therefore, appears to have no independent influence beyond the objectives that are built into it by management. Writers who have adopted the strategic choice approach emphasise the scope for managerial discretion in the way technology is designed, the objectives for which it is used, and the manner in which work is designed around the technology. The scope for strategic choices in all these areas, they argue, suggests that at best technology is of only secondary interest:

> Both technology and its effects are the result of a series of management decisions about the purpose of the organisation and the way in which people should be organised to fulfil that purpose. This implies we should not be studying technology at all, but that we should instead be analysing managers' beliefs, assumptions and decision-making processes.
>
> (Buchanan and Huczynski 1985: 221)

However, there are a number of reasons why this tendency to reject a serious analysis of the independent influence of technology is misguided (see also Clark *et al.* 1988: 9–12, 29–30). First, it is a contradiction in terms to suggest that technology itself is of little significance to the process of technological change. Indeed, if this were true there would be little point in organisations ever introducing new technology. As Winner has argued: 'if [technology] were not determining, it would be of no use and certainly of little interest' (Winner 1977: 75).

The problem with many studies influenced by the labour process and strategic choice perspectives is that in their efforts to avoid what is seen as the technological determinism of writers such as Woodward, the possibility that technology might be *one* of the factors shaping the outcomes of change is ignored. As we have argued elsewhere, there has been a tendency to discard the technology 'baby' with the determinist 'bathwater' (Clark *et al.* 1988: 11).

Nevertheless, while denying the independent influence of technology at an analytical level, much recent empirical research does make passing reference to particular influences that information and computing technologies have had on the outcomes of the change studied. For example, Gallie's investigations in oil refineries revealed that automated process technologies were 'conducive to a degree of team autonomy' (1978: 221). Similarly, the findings of Buchanan and Boddy's Scottish case studies lead them to assert that, 'when technology changes, the tasks that have to be performed change' (1983: 244). In his research on microprocessor-based applications in batch engineering, Wilkinson refers to 'technical constraints' which 'obviously do exist' and place limits on such things as 'the control that workers can exercise over their work' (1983: 66).

In similar vein John Child, who tends to relegate technology to a relatively minor place as a derivative of prior strategic choices, makes the general point that 'a given technological configuration (equipment, knowledge of techniques, etc.) may exhibit short-term rigidities and perhaps indivisibilities, and will to that extent act as a constraint upon the adoption of new workplans' (Child 1972: 6). While terms such as 'conducive', 'constraint', 'place limits on' by no means imply a strong causal or determining influence of technology, they do indicate a recognition of at least some independent role for technical factors in shaping work and organisation.

Of course, if one still insists on the view that the impacts of technology are a direct and unavoidable manifestation of the inherent logic of capitalism, it could be argued that observations such as these are relatively insignificant (see for example, CSE 1980; Albury and Schwartz 1982). However, we would argue that adherence to what is itself a rather mechanistic and economistic determinism can be objected to on a number of grounds (see also Clark *et al.* 1988: 9–12). First, to suggest that because a technology is designed and used within a capitalist mode of production it is necessarily impregnated with capitalist values is not a view that should be accepted *a priori*. In practice, such direct or necessary links between the social, economic, political and technological context of the innovation process need to be established empirically. Unfortunately, few studies of the design and innovation of process technologies have been undertaken with the explicit intention of uncovering the way social and political choices, and in particular those reflecting the 'class interests' of employers and management, have been embodied in particular technologies (for one notable exception see Noble 1984).

Second, even where such social and political influences are found to be evident, it would be wrong to conceive the innovation process as shaped entirely by non-technical factors, since design choices also arise from, and are constrained and extended by, existing technology. As Mackenzie and Wajcman point out, new technology 'typically emerges ... from existing technology, by a process of gradual change to, and new combinations of, that existing technology . . . Existing technology *is* thus ... an important precondition of new technology' (1985: 8–13; original emphasis – see also Roy 1986).

Third, in many cases of process innovation, the technology is designed and manufactured by other firms[1]. As a result, management in the adopting organisation may have little scope to influence directly the basic design features of the technology, although they may be able to set specifications or select from alternative types of equipment or system. In other words, whilst social and political choices may have influenced the design of a particular technology in the supplying firm, it cannot be assumed that this will necessarily reflect managerial interests in the adopting firm. In this way a technology, once purchased from a supplier, can have a significant independent influence on the outcomes of change. As Wilkinson observes, although 'machines may embody particular configurations of power and control', within organisations adopting the technology 'constraints on possible work organisations may already have been *inbuilt* during the design process, which could have been carried on largely externally ... In this sense technology *can* have 'impacts' on work organisation and skills (1983: 21, original emphasis).

Finally, whatever the nature of the social and political influences on the design of a technology, and the reasons for its choice by an adopting firm, this does not automatically mean that the technology itself will necessarily have the intended effects when it is introduced. Suppliers often overstate the capabilities of their products or supply customers with systems that do not in practice perform to the specification given. In particular, in the early stages of the diffusion of an innovation where adopting firms are not experienced with a new technology, many firms are persuaded to buy systems which do not suit their requirements – in short they are 'sold a pup'. In circumstances such as these, a technology, *once chosen*, can again exert a significant independent influence on subsequent choice and negotiation in an adopting firm simply because it does not perform to the specification that management believed it would. Indeed, these latter two points may be applicable particularly in the case of new computing and information technologies where adopting firms are inexperienced with such systems and have little in-house expertise to enable them to evaluate suppliers' claims.

To summarise, we would argue that a preoccupation with avoiding 'technological determinism' has meant that the need for a detailed analysis of the independent influence of technology itself – although evident in the observations made in a number of empirical studies – has not been realised in practice. This, we suggest, is the result of a confusion between the determinist assumption that technology is the *primary* influence on the outcomes of change, and the rather different point that technology can have an *independent* influence on outcomes. In this sense, Woodward's work is of considerable value since she does attempt to develop an analytical framework, however flawed, which takes into account the independent influence of 'technology'.

Analysing the independent influence of technology

An immediate problem which confronts any attempt to analyse the independent influence of a technology is the question of how 'technology' itself is to be defined. Although the answer to this question may at first appear simple there is in fact little or no consensus on the best approach to adopt. This section, therefore, discusses existing definitions and their weaknesses and outlines an alternative approach which in the subsequent section is used to analyse the independent influence of two examples of computing and information technology on work tasks and skills. Finally, an attempt is made to see how far these findings are generalisable to other types of new technology.

Approaches to defining 'technology'

Much of the disagreement that has arisen about the actual influence of technology derives from the wide variety of definitions that have been employed by researchers (see Winner 1977; also Buchanan and Huczynski 1985: 211–14 for discussion). Some writers have adopted a highly restrictive notion based on the idea of technology as equipment or apparatus (see for example, Rice 1958: 4; Pugh and Hickson 1976: 93; Thompson 1983: 9; Child 1984: 38; Batstone *et al.* 1987: 2). Others, including Woodward, have also incorporated within their definitions the techniques and principles governing the way work operating the equipment is performed. Woodward's definition in fact embraces both the 'collection of plant, machines, tools and recipes available at a given time for the execution of the production task' and 'the rationale underlying their utilisation'. (Reeves, Turner and Woodward 1970: 4). Similar in this respect is Rose who adds to the confusion by labelling the physical items related to the task of technical production as 'hardware' and the work roles and division of labour which surrounds it as 'software'! (Rose 1978: 138).

Some definitions have gone so far as to embrace the organisational features of the social arrangements behind the use of technology. Fox, for example, distinguishes between 'material' and 'social' technology where the former refers to equipment and the latter to the social organisation of work through structures of co-ordination, control, motivation and reward systems in the form of job definitions, pay structures, authority relations, systems of communication and control, and disciplinary codes and rules (1985: 1).

Given this plethora of definitions it is no wonder that one commentator should conclude that '... research dealing with the influence of technology on structure is not only conflicting, but in extensive disarray' (Bedeian 1980: 234). The problems with existing approaches to defining 'technology' are twofold. First, they all share a tendency to treat the hardware and software which comprises technology as a 'black box'. By definition therefore they do not seek to analyse in detail the technical characteristics or capabilities of the equipment itself. Second, by collapsing together such things as hardware, software, techniques and principles, skills, knowledge and organisational arrangements, in one all-embracing definition of 'technology', the broader approaches cause considerable problems when used as analytical tools. For example, by failing to distinguish between hardware and software on the one hand and the skills and knowledge required to use them on the other, it is virtually impossible to make any meaningful statements about the relationship between the changes in work tasks and skills which occur when technology changes. As Buchanan and Huczynski note, in so far as changes in work tasks and skills can be regarded as dependent upon changes in hardware and software, this leads to the confusing situation where the independent variable, the equipment, overlaps with the dependent variable, the techniques used to perform work using it (1985: 211–12).

The first step to resolving the analytical problem of how to define technology is to decide whether to adopt a restrictive or expansive definition. Winner suggests adopting a definition of technology which recognises three separate dimensions: 'apparatus', referring to the physical equipment itself; 'technique', referring to the skills and knowledge required to use the equipment; and 'organisation', referring to systems and structures of control and co-ordination (Winner 1977). While at least having the merit of recognising that technology can be defined at a number of levels, Winner's approach clarifies the options but does not resolve the ambiguity and confusion already noted. The remainder of this section outlines an alternative definition.

Opening the 'black box': technology as engineering system

This section draws on attempts by ourselves and colleagues to develop a more sophisticated definition of 'technology' for use in empirical research (see Rose *et al.* 1986; Clark *et al.* 1988). The approach taken involves a restrictive definition in the sense that 'technology' is taken to refer to equipment and apparatus (including both hardware and software), but also draws on engineering concepts in order to avoid treating the equipment and its capabilities as a closed 'black box'. Our approach seeks to 'open the black box' by defining 'technologies' as *engineering systems*. That is, rather than just being pieces of hardware and software, 'technologies' are also conceptualised as systems based on certain engineering principles and composed of elements which are functionally arranged (configured) in certain specific ways. In this way 'technologies', or more accurately 'engineering systems', can be defined in terms of three primary elements '. . . system principles, an overall system configuration, and a system implementation or physical realisation in a given technology'. The first two elements can be termed architecture, and the third technology (Clark *et al.* 1988: 13).

Engineering systems are also characterised by two further elements, which are largely derivative of system architecture and technology. The first refers to the way a system is tailored or dimensioned to suit the physical layout of a particular workplace or customised to meet the requirements of a particular user application. For example, whilst CAD systems constitute basic types of 'engineering system', they are likely to be *dimensioned* in specific applications to provide a certain number and layout of work stations and to have applications software customised to meet the specific user's needs. The second element refers primarily to the visual and audible characteristics of the technology of an engineering system. This system appearance can be particularly significant where a change from mechanically or electro-mechanically based to electronics or microelectronics-based engineering systems is involved, and can also have a direct bearing on the content and experience of work.

The concept of *engineering system* is presented schematically in Figure 5.1. The advantage of this definition is that it allows for a detailed analysis of the precise links between aspects of a particular technology (engineering system) and a range of variables such as work tasks, skills, job content and supervision. It does not mean necessarily that changes in technology (engineering systems) determine changes in these other variables, but it does imply that it may be one factor influencing (enabling or constraining) the design space available for the social choices during the process of change. In other words, the approach allows for an analysis of the independent influence of technology which 'is not a replacement but a complement to the analysis of social variables' (Clark *et al.* 1988: 32).

Figure 5.1   The concept of engineering system

PRIMARY ELEMENTS	
Architecture	*Technology*
system principles	hardware
overall system configuration	software
SECONDARY ELEMENTS	
Dimensioning	
detailed design for a particular	
organisational setting	
Appearance	
audible and visual characteristics	
ergonomics	
aesthetics	

Source: Clark *et al.* (1988)

New technology, work tasks and skills

It is now possible to analyse the independent influence of new computing and information technologies on work tasks and skills. In order to accomplish this, detailed evidence from two studies in which we have been involved will be examined. The first involves the replacement of electro-mechanical by semi-electronic telephone exchange systems in British Telecom. The second concerned the replacement of manual drawing boards and equipment by CAD systems in engineering drawing offices. Finally, the influence of these technologies on task and skill requirements will be compared with effects documented in research concerning other examples of computing and information technology.

Modernisation of telephone exchanges

In this case, the technological change in question was the introduction, during the late 1970s and early 1980s, of a semi-electronic telephone exchange system known as 'TXE4'. This took place as part of an overall plan to modernise the telecommunications network and replace the old electro-mechanical Strowger system (for a full account of what follows see Clark *et al.* 1988). The research revealed that the new TXE4 system had an independent influence in a number of ways. The most direct influence was found in relation to maintenance work tasks and skills. The characteristics and capabilities of the system had identifiable but less strong implications for the way maintenance skills were acquired and the content of the maintenance technicians' jobs. Finally, the design space available for devising ways of supervising and controlling maintenance work was extended and constrained in important, if not necessarily always decisive ways. What follows will pay particular attention to the relationship between changes in engineering systems and changes in the task and skill requirements of maintenance technicians.

The function of a telephone exchange is to connect or 'switch' calls. The earlier types of telephone exchange required a human operator to perform the task of connection manually by means of a plug and socket. However, by the end of 1976 all exchanges in Britain had been automated, thereby eliminating the need for human intervention in the switching of calls. Automated exchange systems are able to recognise that a customer wants to make a call, interpret the destination instructions and make the necessary connection. Exchange systems thus have two main capabilities; they can accomplish the physical connection or *switching* of calls, and they can *control* the process of setting-up, sustaining, and terminating calls. As with other automated processes, however, 'indirect' labour is required to perform maintenance. The major motivation in the design of exchange systems based on electronics and microelectronics technologies has been to improve the quality and range of services available to the customer, while reducing the cost of maintaining exchanges. As such, all the features of electronics technology outlined in Chapter 1 – speed, processing power, reliability, flexibility, reduced size and cost – are all important in the design and operation of telephone exchanges.

The Strowger exchange was invented in the late nineteenth century by a Kansas City undertaker, Almon Brown Strowger. He suspected that the wife of a competitor working as a local telephone operator was re-directing customers phoning Mr Strowger to her husband! This served as a strong motivation for Strowger to invent the first fully automatic telephone exchange which replaced the need for a human operator to connect calls manually. Exchanges based on Strowger's design were first introduced in Britain during the 1920s, and in the early 1980s were still responsible for around fifty-five per cent of all exchange connections.

The Strowger system is based on a telephony principle known as 'step-by-step' which involves a succession of switching stages controlled directly by the dialling action of the person making the call. As a number is dialled by the caller this produces electrical pulses which activate, in a 'step-by-step' fashion, a series of switches to set up and sustain a connection (see Figure 5.2). When one of the connected parties picks up his or her receiver the connection is made. Replacing either receiver breaks the connection. The switches for each stage of the switching process are arranged (configured) in successive racks to minimise the length of interconnecting wiring. In principle, this system configuration makes it possible to trace the electrical path of a call through the successive stages of switching equipment.

The exchange architecture is implemented using electro-mechanical technology. The Strowger switch comprises electro-magnetic relays which control the movement of a rotary contact arm, responding in part to the electrical pulses generated by the dialling procedure. Strowger exchanges also have distinctive visual and audible properties or system appearance. Switches are mounted on ceiling-to-floor racks, and the moving parts of the switch can be seen as they move up and down and rotate in order to set up and terminate calls. This action is audible as a 'staccato rattle' made by the enormous number of moving mechanical parts in the

Figure 5.2   The Strowger step-by-step switching system

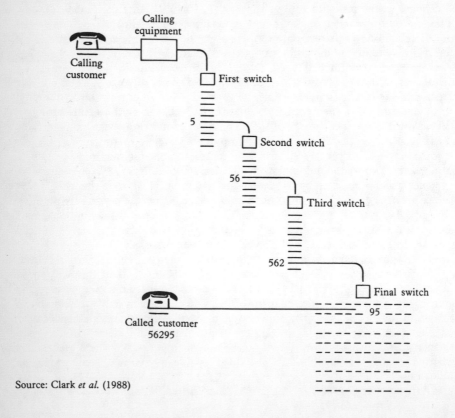

Source: Clark *et al.* (1988)

exchange. The principal feature of Strowger exchanges is that they require a high level of costly routine preventive maintenance to stop the switches becoming dirty, jamming and even breaking – all faults which affect the quality of service to the customer.

The TXE4 system was introduced into the exchange network as a replacement for medium to large-sized local Strowger exchanges. In contrast to the Strowger exchange, the architecture of the new TXE4 system employs telephony principles known as 'common control' and 'matrix switching'. The control function is implemented using electronic technology and the switching function electro-mechanical technology. Unlike Strowger exchanges where the control and switching function is combined in each switch, TXE4 exchange architecture involves a separation of the control elements from the switches, concentrating them in one functional area of the exchange. This equipment is then shared or 'common' to all switching operations. The 'common control' equipment uses electronic components and is similar to a standard computer. The computer acts as the 'brain' of the exchange, memorising the wanted telephone numbers as they are dialled by callers and routing calls through the exchange to make connections – all within fractions of a second.

The switching function utilises a 'matrix switching' principle (see Figure 5.3). This simply means that a call is set up by operating cross points on a series of matrices to connect the caller with the desired party. Connections are made by activating electro-mechanical switches, different from and more reliable than Strowger switches (see Figure 5.4). Compared with Strowger exchanges, the range of routes a call can take through a TXE4 exchange is far wider. Should a fault occur which prevents a connection being made, the computer controlling the exchange is able, in fractions of a second, to make further attempts to connect the call which are unbeknown to the caller. The computer also captures and displays information on fault conditions on a teleprinter.

Figure 5.3   The matrix switching principle

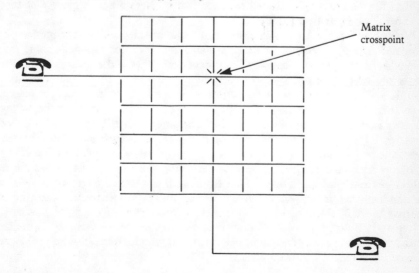

Figure 5.4　TXE4 matrix switching

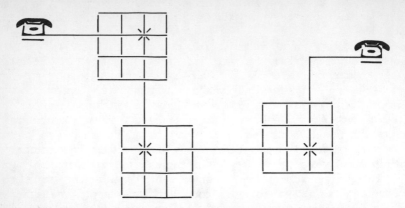

Source: Clark *et al.* (1988)

One important feature of the common control architecture is that it involves a far greater functional interdependence between parts of the exchange, with failures in one area of the exchange system having repercussions elsewhere. *In extremis* there is the possibility that the whole or part of the system could crash 'off the air' – something much less likely in a Strowger exchange. The system appearance of TXE4 exchanges is also radically different to Strowger. Both electronic and electro-mechanical components are mounted on slide-in cards, and neither the operation of the control equipment nor the activation of the switching process is observable, although the latter is audible as a faint 'clicking' sound. The principal feature of the maintenance requirement of TXE4 exchanges is that their architecture makes them more 'fault tolerant' than Strowger exchanges, while their implementation using more reliable technology reduces the requirement for routine preventive maintenance. In principle, therefore, the system offers a better quality of service while being cheaper to maintain.

Although the control and switching functions of telephone exchanges are fully automated, the maintenance of these systems still requires human labour. At the heart of most maintenance tasks in telephone exchanges is the process of 'faulting', that is the identification, location, diagnosis and repair of fault conditions (for example where a connection fails or where a crossed line occurs). In our research we found that the faulting task was shaped in three principle ways by the Strowger exchange system. The step-by-step architecture meant that it was possible to locate faults by literally following the path of a call through each successive switching stage. The co-location of control and switching functions in each switch meant that faults could be isolated, for example by removing the faulty switch from the rack, without normally affecting other parts of the exchange. Exchange architecture also had implications for the skill required. For example, performance of faulting tasks required 'mental' skills based on a theoretical knowledge of the principles of step-by-step switching in order to be able to follow the path of a call through the exchange, and an understanding of the theory on which electro-mechanical systems are based (for example Ohm's Law).

The electro-mechanical technology, and its requirement for regular routine maintenance, also had a direct and distinctive influence on many work tasks. The equipment had a tendency to accumulate dirt, become misaligned, jam and even break. Repair and routine maintenance tasks, therefore, required a range of manual skills and abilities in cleaning, lubricating and adjusting the equipment. Finally, system appearance had an important influence on fault location and diagnosis, since faulty switches could be visually and aurally identified (faulty switches sounded different and very occasionally would make their presence visually obvious by ejecting components on to the exchange floor!). This called for refined aural and perceptual abilities. Finally, underlying the performance of maintenance tasks in general were tacit intuitive and experiential skills which were a mixture of accumulated 'local knowledge', 'inspired guesswork', and manual dexterities of 'touch' and 'feel' (see Clark *et al.* 1988: 91–100).

The architecture and technology of the new TXE4 exchanges generated radically different task and skill requirements when compared to Strowger. Compared to Strowger, there was a greater amount of information to assist faulting from automatic test equipment and the exchanges' own in-built, self-diagnostic capabilities. However, the functional interdependencies of the exchange equipment and the wide range of routes a call could take through the exchange meant that this information was only a broad guide to locating and clearing the fault. The implementation of the control function of the exchange using electronic technology meant that some faults could be highly complex and their cause remote from the symptoms. However, repair of an electronic component was usually a simple task of replacing a slide-in card or soldering on a new component to replace a faulty one.

There was therefore a requirement for a different order of mental diagnostic and interpretive ability, combined with a broader system understanding compared to Strowger, in particular since the TXE4's system appearance gave no direct visual or audible clues to its operation or fault conditions. One technician described faulting in TXE4 as a process of 'chasing rainbows' and of going down many 'blind alleys' in order to locate a fault. This prompted the description of the operating principle behind TXE4 as 'Sod's Law' (in contrast to the use of Ohm's Law to solve electro-mechanical faults in Strowger exchanges!).

The new skills required to accomplish TXE4 maintenance tasks showed a qualitative change in contrast to Strowger. On the one hand the electro-mechanical theory, manual dexterities and elements of tacit knowledge associated with electro-mechanical systems were no longer needed. On the other hand there was now a strong emphasis on mental diagnostic skills. There was a need to think things through rather than 'dive-in' to solve problems by trial and error. Problem-solving involved a clear understanding of electronics and the 'common control' principle in order to be able to conceptualise the exchange as an interdependent system and to interpret information from test equipment and computer print-outs. As one maintenance technician observed:

> . . . you've got to know a lot more. You've got to take the whole exchange in context rather than just one little piece of equipment . . . In . . . (Strowger) . . . you could isolate and work away on one piece (of equipment). On (TXE4) you've really got to think of what's happening all through the (system). A different kind of thinking what the problems are.
>
> (Quoted from Clark *et al.* 1988: 116)

The architecture and technology of the new electronic exchanges, therefore, had a direct influence on changes in task and skill requirements, removing the need for many manual skills and abilities but requiring new problem-solving mental skills. As a result of its more abstract content, the relationship between the technicians and their work on the exchange system became less direct. The technicians made frequent references to Strowger as a 'hands on' system while TXE4 required a 'hands off' approach.

In conclusion, the independent influence of the new technology on skills did not involve either a uniform tendency towards a 'deskilling' or 'upskilling' of work tasks. Rather both processes were in evidence at the same time. On the one hand 'deskilling' occurred in so far as the manual skills associated with Strowger work tasks were no longer required; on the other, 'upskilling' occurred in the form of a requirement for a qualitatively different level of mental diagnostic and system skills.

Computer-aided design systems

The second example of the effects of new technology on work tasks and skills is provided by the replacement of conventional drawing boards and manual drawing implements with computer-aided design (CAD) systems. The effects of these systems were studied in four firms by one of the present authors (see McLoughlin 1986; 1987; 1988). This revealed a similar shift from manual task and skill requirements to mental problem-solving abilities. CAD systems usually consist of some form of input device for information capture, a central processing unit and magnetic discs for information storage, a visual display screen for the display and manipulation of graphics through various input devices, and a printer and plotter for information distribution (see Figure 5.5). A range of applications software enables numerous drafting and detailing operations to be controlled via the computer. More sophisticated systems allow the convergence of a variety of design and drafting tasks so that they can be accomplished using one system. The most advanced systems offer the possibility of increased integration between the design and manufacturing process by enabling the communication of design information in the form of digital output to be made direct to the shop floor – thus reducing the need for engineering drawings.

The most immediate implication of the CAD systems for work tasks and skills revealed by the research derived from their implementation using technologies which are radically different to traditional manual tools. Rather than drawings being made directly on to paper using a pencil, T-squares and so on, drawings were actually made by using electronic input devices. Usually these took the form of a 'digitising tablet' which contained a 'menu' that could be used to call on to the screen standard parts or components. A 'puc' (similar to the 'mouse' now provided with some personal computers) was used for activating routines on the menu tablet and for drawing lines on the screen. The task of drawing using this technology required different skills to those associated with manual methods. For example, the task of drawing a tangent to a circle by conventional means is simply a matter of positioning a ruler and drawing the line with a pencil. The same operation using CAD involves manipulating a 'puc' to make several inputs on the digitising tablet in order that the line is drawn automatically on screen in the desired position (Jonas 1988).

Figure 5.5   'Turnkey' CAD system – typical hardware configuration

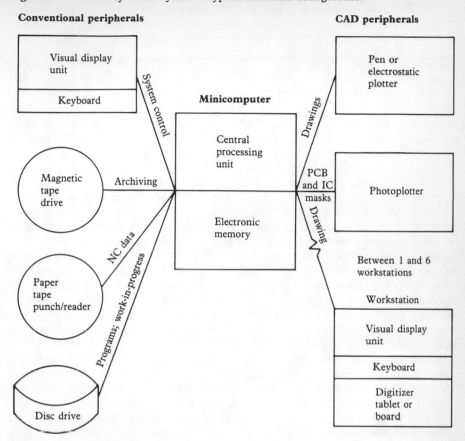

Source: Arnold and Senker (1982)

Work tasks involving drawing by conventional means utilised a number of craft manual skills in manipulating a pencil and required a direct relationship between the thought of the user to 'draw a line' and the act of drawing. Work tasks involving drawing with CAD required more simple manual abilities to move the 'puc' and sometimes to input data through a keyboard. The manual skills used in actually drawing were eliminated, as lines of perfect quality and weight were drawn automatically by the system. The relationship between the user and the drawing thus became more indirect and abstract, requiring the exercise of mental skills in understanding the capabilities of the system and in selecting appropriate routines from the menu. This difference was expressed in the following terms by one manager in charge of a CAD facility in a building firm:

On a drawing board you're thinking construction all the time. The pen just flows, you move your hand around. You've got one train of thought, how you construct, how you design. With CAD you've got to think of construction, of how to do it using the software and the best way to do it, because there are always six ways of doing it, you haven't got that option using the board. You just draw it. There's more thought processes going on with CAD.

(Quoted from McLoughlin 1986: 22)

Many writers suggest that the loss of manual craft skills in drawing tasks is positive evidence of deskilling (see for example Cooley 1980; 1981; Kaplinsky 1982; Baldry and Connolly 1986; Jonas 1988). However, this is by no means clear cut. In the cases studied the new mental skills required in using CAD did in some ways compensate for this loss, and the kind of personal touch associated with manual methods, for example in lettering and layout, was still required (see McLoughlin 1986; 1988).

However, the major influence of CAD on work tasks and skill requirements derived not so much from the technology used, but from the architecture of the systems. Broadly speaking, CAD systems are of two types: *drafting systems* and *modelling systems* (Taber-Heyder 1985). Drafting systems are in effect electronic drawing boards which enable the user to manipulate a two-dimensional drawing comprising lines and curves and annotated by figures and characters. The capabilities of drafting systems are essentially those of a 'shape processor' – an analogy with word processing. As with word processors, drafting systems offer advantages over manual methods in the ease and speed of correction, the insertion of new material and amendments, and in the production of repetitive material. In addition, drafting systems offer additional facilities such as the manipulation and display of complex drawings in 'layers'[2], and more accurate dimensioning, which on most systems can be accomplished automatically.

In contrast, modelling systems set up an 'electronic' model of a three-dimensional object which is stored in a computer database and from which drawings can be extracted and displayed. These can then be manipulated as with 'drafting systems', with the added advantage that any changes made are updated on the model, which also updates other user files as appropriate. There is thus a far greater functional interdependence between the activities of the system users, since the consequences of changes made by one user can immediately be seen for others. Moreover, there are also possibilities for greater integration of design with manufacturing systems as the output from modelling systems can be used 'downstream' in the production process, for example, in the programming of computer-aided manufacturing technologies such as CNC machine tools, and 'upstream' in the evaluation of different design options.

The implications of CAD for drafting and design tasks and skills in the cases studied were numerous. Three of the four firms used CAD 'drafting' systems. In these cases, whatever the changes in the skills needed to 'drive' the system, engineering knowledge, expertise and experience were still required. This point was illustrated in one case – an engineering consultancy – by the following anecdote. During a particularly busy period one of the CAD operators who specialised in mechanical engineering drawings was asked to help out by revising a batch of architectural drawings. The results were received amid great hilarity in the drawing office when it was seen that the individual concerned had detailed every brick on the elevations of the building depicted in the drawing. Whilst customary in mechanical engineering to mark the location of every bolt, there was not an equivalent requirement in architectural drawing! Some commentators have suggested that engineering knowledge and expertise could be designed into CAD software thus making it possible for a 'non-engineer' to operate a CAD terminal. However, such software requires a high degree of time-consuming customisation to be applied to even the most routine of detail drawing tasks. In none of the cases where drafting systems were in use was this seen as a realistic possibility.

In fact, the skill requirements of work tasks which involve the use of CAD systems with modelling capabilities suggest the need for an enlargement of engineering skills and expertise. One characteristic of the development of the division of labour within the design and drawing functions of many firms has been a specialisation between design and drafting and also a division of labour according to engineering discipline (see Arnold and Senker 1982). For example, professional engineers have become predominantly responsible for creative design, while routine drafting has become the activity of non-professional drawing-office staff. Similarly, electrical engineers have concentrated on electrical and electronic drawings and mechanical engineers on mechanical drawings.

However, CAD 'modelling' systems may challenge this traditional division of labour and engineering expertise (see Winstanley and Francis 1987). This was illustrated by the one case in the four firms studied where this type of system had been adopted. Here, the functional interdependence created between the activities of users of CAD modelling systems, and possibilities for greater integration with design and manufacture, meant that a far wider range of 'composite' engineering skills and knowledge was required in the drawing office. CAD users had to develop an interdisciplinary approach to detailed design so that they understood the implications of their activities for work being carried out on the 'model' by other system users (the 'model' in this instance was a design of a new naval vessel). In addition, the system allowed drafting staff to make more creative design decisions, thus blurring the distinction between them and professional design engineers.

Other research

Other research has not analysed the independent influence of new technology in the detail allowed by using the concept of *engineering system*. Nevertheless, similar influences to those we have identified in telephone exchanges and CAD systems have been noted in so far as manual task and skill requirements are reduced; the need for mental problem-solving skills is increased; a reliance on tacit and experiential skills is still important; and the relationship between the operators and their work becomes more abstract.

For example, in the case of word processing systems it has been found that the need for manual skills associated with the use of conventional typewriters, such as an even 'touch' on the keyboard to get a consistent weight of character when typed, are eroded. Similarly, many aspects of page layout and making corrections are automated by the system. However, new skills and knowledge are also required. For example, computer codes and menus for formatting and editing texts have to be learnt and their meaning when displayed on screen understood. Similarly, knowledge of how to manage files is essential in order to avoid unwittingly erasing work (see Buchanan and Boddy 1983: 161–2).

The need for new mental and problem-solving skills deriving from the system interdependencies and operations control capabilities of computing and information technology has also been identified. In the case of CAD/CAM systems Buchanan found that, while automating activities previously performed manually, the 'systemic' character of these technologies meant that requirements for skilled human intervention in problem-solving increased because 'in such integrated systems, errors go further faster and stand less chance of detection before they arrive on the shop floor' (1985a: 22–3). This view is supported by Francis *et al.* who concluded from their case-study research that with CAD/CAM, 'the experience of work at all levels in the organisation will increasingly be that of managing and coping with complex relationships and problem-solving in collaboration with others' (1982: 193).

The importance of the operator's ability to detect error has also been noted in studies of automated process plants. For example, Gallie found that in oil refineries the volatility of operating conditions and input products required human operators to be able to adjust controls continually in order to keep to required specifications and to be diligent in avoiding errors when opening valves to move processed products around the plant. In particular, because of the high degree of integration between different units, errors were particularly serious at the critical points when plants were started-up or brought to a halt as 'an error on one unit had immediate implications for many others' (1978: 213–14). Similarly, Buchanan and Boddy have noted that process-control operators in a highly automated chemical plant required a high degree of skill in interpreting alarm conditions generated by the computer controls and in deciding on the appropriate course of corrective action (1983: 224).

A number of writers have emphasised that whilst manual skills may be eroded, the new technology does not, however, replace the need for tacit and experiential skills. These are acquired by working 'on the job' and through familiarity with particular processes, products, raw materials, or equipment. Indeed, in many circumstances the effective use of computing and information technologies appears to be dependent upon human operators possessing tacit knowledge and skills (see for example, Noble 1979; Jones 1982; Wilkinson 1983; Buchanan 1985a). The phenomenon has been reported widely in the case of CNC machine tools. Here, although the need for manual craft skills in operating the machine is eroded, experiential and intuitive abilities based on years of experience of, for example variations in raw materials, are still required to deal effectively with operating contingencies which are not anticipated by the programs controlling the machine. As will be seen in the next chapter, this continuing requirement for workers to exercise tacit knowledge and skills can have an important influence on the way outcomes of change are chosen and negotiated.

Finally, the indirect relationship between workers and machines which derives from the more abstract content of work tasks using electronics- and microelectronics-based technologies has been noted by Cynthia Cockburn in her study of the introduction of new technology in the composing rooms of four newspapers. She provides a graphic account of how the new 'cold type' computer technology transforms the composition stage of newspaper production and apparently reduces the requirement for traditional manual craft skills[3]. A major element of composition involves typesetting. Until the 1970s typesetting was accomplished using 'hot-metal' methods based around large Linotype machines, which were typically seven feet high and six feet across with numerous moving parts. Here compositors or 'comps' would receive type-written copy from the editorial department to be typeset:

> The operator sits at a wide keyboard on which 90 keys are organised in three banks . . . each keystroke he makes releases a small brass matrix, a tiny hollow mould of a single letter of the alphabet, from the overhead magazine, the matrices slide one at a time, down the chute in response to the operator's key strokes, collecting in an assembler where they may read the line . . . The collected matrices are then 'sent away' by pressing a lever. Molten metal is forced into the faces of the characters resulting in a solid slug or 'line o' type' about one inch high, which is ejected on to a waiting tray (galley). Here, the lines, still hot to the touch, assemble, as the operator taps away, into columns of text.
>
> (Cockburn 1983: 47–8)

The replacement of 'hot metal' by computer-based methods of typesetting radically changed the experience of work for compositors:

> The men . . . report a striking change in their relationship to the equipment on which they work. The Linotype was large, its parts were visible and moved . . ., the men knew the function of each component, they listened for changes in the sounds made by the machine and would respond to them . . . The new electronic keyboards however are small, smooth, encased and unrevealing . . . Most of the men had had a glimpse inside the input unit. They saw an enigma. 'There's nothing moving in the damn thing. It's all chips and solder.' Men brought up in a mechanical era, used to cars as well as Linotype feel helpless before computer technology . . . In such ways the men have moved from an active and interactive relationship to a passive and subordinate one.
>
> (Cockburn 1983: 102)

This finding shares an affinity with those noted above in the change from the visible and audible electro-mechanical Strowger exchange to the more indirect and abstract work content in TXE4 exchanges. However, here the work relationship was not one of subordination to the technology. This suggests the possible influence of social choice and negotiation in the design of work since, as will be seen in Chapter 6, the skill content of jobs is shaped not only by task and skill requirements but also by the way work is designed.

In Chapter 1 it was noted that the information handling and operations control capabilities of computing and information technologies do not involve the elimination of all human labour in the control of work in any literal sense. In reality these systems tend to render work more automated than it was previously. This is an important point since it helps us to understand why new task and skill requirements are generated by the technology. As Buchanan and Boddy note, information storage, manipulation and distribution take place with high degrees of reliability and predictability inside the equipment, the only human involvement needed being to activate the procedures. However, information capture, the receipt and interpretation of distributed information, operations control and maintenance, do still normally require human intervention (1983: 12).

Indeed, computer-generated information can complement the interpretive skills and abilities of humans. Whilst new technologies do further erode requirements for humans to execute tasks concerned with the transformation and transfer of raw materials, they do not as the literal interpretation of the term 'automation' implies reduce the requirement for informed human involvement altogether. Rather, new computing and information technologies can require increases in informed human intervention to interpret information, monitor the process and to be able when required to make decisions and solve problems. As Buchanan and Boddy observe, there is:

> ... a continuing need for human presence and informed intervention in the effective use of computing and information technologies. They replace physical effort and manual skills, increase the rate at which physical operations can be carried out, improve consistency in work flow and output, and generate information that can be used by human operators to improve control. Their information handling and control capabilities complement, rather than replace, human information processing, problem solving and decision-making skills.

> (1983: 236)

Chapters 6 and 7 will explore some of the implications of these new task and skill requirements in the context of choice and negotiation over the design, supervision and control of work around new technology.

Conclusion

This chapter has been concerned to identify the independent influence of new computing and information technologies on work tasks and skills. In so doing it has gone against the grain of much recent research influenced by labour process and strategic choice approaches which has played down the significance of technology in order to avoid what is seen as 'technological determinism'. We have argued that, in effect, this results in the technology baby being thrown out with the determinist bath water. Attention was then turned to identifying the specific influences of new computing and information technologies on work tasks and skills. In order to carry out this analysis it was necessary to develop a more sophisticated analytical definition of technology in terms of the concept of *engineering system*. This definition was then used to identify the influences of two types of new technology (semi-electronic telephone exchanges and CAD systems) on task and skill requirements. The conclusions drawn appeared to be supported by research on other types of computing and information technology.

The broad conclusion of this chapter therefore is that these technologies may generate imperatives which have the following effects on work tasks and skills. First, they eliminate or reduce the amount of complex tasks requiring manual skills and abilities; second, they generate more complex tasks which require mental problem-solving and interpretive skills and abilities and an understanding of system interdependencies; third, in order that many tasks can be performed effectively tacit skills and abilities associated with the performance of work with the old technology are still required; fourth, they involve a fundamentally different relationship between the user and the technology compared to mechanical and electro-mechanical technologies.

Notes

1 The PSI survey, for example, found that only three per cent of manufacturing firms adopting microelectronics-based process innovations had designed and made the system in their own factory. In eight per cent of cases the system had been made to the company's specification by an outside sub-contractor, in ten per cent of cases it had been specifically designed outside the company, and in sixty-seven per cent of cases the system was a standard catalogue item from an outside supplier (Northcott 1986: 53). Moreover, as shown in Figure A (see Appendix) by far the most frequently cited disadvantage and difficulty in adopting microelectronics-based systems in processes was the absence of people in the organisation with microelectronics expertise.
2 The 'layering' of drawings refers to the capability of a CAD system to separate out component elements of a drawing and overlay them one on top of the other. The effect is analogous to the process of making up a drawing with elements drawn on separate sheets of tracing paper and then placed one on top of the other. CAD systems can allow drawings to be composed in a large number of separate layers. This is a particularly powerful tool where complex drawings involving large numbers of lines are involved, or where it is necessary to separate out one element of a drawing – for example the electrical wiring in a building or a ship – from all the other components for a particular presentation.
3 The composition stage of newspaper production is the point at which copy generated by the editorial and advertising departments is 'made-up' into pages. Once the pages are made-up or 'composed' they are then ready to pass on to the next stage for printing. For a full description of the stages of newspaper production see Martin (1981).

CHAPTER 6

New technology, job content and work organisation

Chapter 5 explored the independent influence of new technology on changes in work tasks and skills. This analysis was conceived as a complement to and not as a replacement for an examination of the way the outcomes of technological change are chosen and negotiated by managers, unions and workforce. In Chapter 5 we suggested that when a new technology is introduced into the production process the work tasks that have to be accomplished also change. However, this still leaves open important questions of whether and how these new work tasks are to be allocated to existing jobs or grouped into new ones, and how these jobs are to be linked into a more general pattern of work organisation and supervision. These 'control issues' are the concern of this and the next chapter. This chapter begins with a consideration of managerial choices in the design of jobs around new technology, and goes on to assess the formal and informal influence of unions and workforce in negotiating and re-negotiating working arrangements and practices during and after the process of change. As will be seen, the choices made by managers and the interventions of trade unions and workforce can have an important bearing on the outcomes of technological change – in particular in relation to skills.

Work tasks, job content and skill

The content of jobs is usually taken to refer to a variety of dimensions which go together to make up the work that workers and workgroups carry out in the workplace. The most frequently cited dimensions relate to the *work tasks* involved in a particular job – including such things as the range and variety of tasks, the complexity of tasks, and the effort levels required to accomplish them – and the *discretion* or *autonomy* that the job-holder has in executing tasks – which can include the various controls and sanctions under which the worker or work group operates (see Bailey 1983; Child 1984; Batstone *et al.* 1987; Rolfe 1987; Buchanan, forthcoming). A key aspect of job content is the skills required by job-holders – what can be termed the 'skill content' of work. Heather Rolfe has suggested that the skill content of work can be defined in terms of the *technical complexity* of jobs – measured in terms of task complexity, knowledge requirements and task range and variety – and, the amount of *discretion* involved – measured in terms of the degree of decision-making and judgement required, the extent of control over the organisation of work and the limits imposed by supervision (Rolfe 1987: 86–92). How then may managers and unions seek to influence job content and work organisation?

The evidence discussed in Chapter 3 on management strategies for introducing new technology pointed to the lack of importance attached by managers in the early stages of change to decisions over job content and the pattern of work organisation. Where they are considered at all, these questions of what Buchanan (forthcoming) terms 'work design' appear to be left for line managers to determine at later stages in the process of change, when new equipment and systems are being installed or even after they become operational. It was also suggested in Chapter 3 that lower-level managers tend to see new technology as a means of increasing the predictability, consistency, orderliness and reliability of work operations and of reducing uncertainty by allowing them to rely more on technology and less on informed human intervention. For many writers, especially those sympathetic to the labour process perspective, such objectives are often viewed as an expression of Taylorist assumptions and strategies which aim to deskill work. However, we argued that attempts to reduce the requirement for skilled labour in the production process are not necessarily a manifestation of Taylorism or its derivatives. Rather, it was suggested that management attempts to increase control over labour may in fact be a function of the pursuit of what we termed operational control objectives concerned with the overall performance of the production process.

In the case of trade unions, it was seen in Chapter 4 that the traditional approach to negotiating over technological change has been to focus on issues relating to terms and conditions of employment. Questions of job content and work organisation, along with strategic issues relating to investment and the choice and design of new technology, have not been seen as primary concerns for collective bargaining. In theory at least, the ill-fated new technology agreement initiative was intended to extend the influence of collective bargaining over new computing and information technologies into such areas. In practice, even where trade unions have been relatively successful in negotiating the introduction of new technology, the focus has still been very much on employment issues.

The remainder of this chapter will be concerned to explore a variety of case study and survey evidence in an attempt to establish the relative influence of managers, unions and in particular workgroups on the main 'control issues' associated with the introduction of new technology.

Management influence on job content and work organisation

According to Buchanan and Boddy, 'the changes to job characteristics that accompany technological change reflect partly the capabilities of the technology, and partly the objectives and expectations of management' (1983: 246). This section draws on a range of case-study evidence to illustrate how in practice managerial decisions have influenced job content and work organisation. One important conclusion is that where the new technology has the capacity to complement rather than replace human problem-solving and decision-making skills, decisions to design work in a way which seeks to reduce management's reliance on skilled and informed human intervention can undermine the effective use of these systems.

Computerised controls on the shop floor

In a study of the introduction of computerised process controls at United Biscuits, Buchanan and Boddy (1983: 173–202; also Boddy and Buchanan 1986) provide an example of how management choices can vary within one company with different consequences for job content and work organisation. In one instance this led to work being designed with the intention of replacing the need for human intervention and in another to new technology being used to complement human skills and abilities. These choices were both made in the context of strategic objectives which aimed to maintain profit margins by controlling costs and investing in new technology. In fact, the emphasis placed on investment by the chairman of the company influenced management thinking throughout the organisation. This gave rise to plans to computerise the control of the biscuit-making process. However, within this overall policy framework individual managers had considerable freedom over how to introduce new technology, and hence how to design work.

The process of biscuit making has three stages: mixing, baking and wrapping and packing. At the mixing stage, computer controls were introduced which replaced the mixing of dough in open vats by manual workers under the supervision of time-served master bakers called 'doughmen'. The computer now controlled the adding of ingredients to the vats automatically. Recipes for biscuits were contained on punched paper tape and fed into the computer by operators in a control room. The punched paper tape method was later replaced by a microelectronics-based 'recipe desk' which gave the control-room operators greater flexibility in adjusting recipes. The doughmen's jobs were redesigned so that they were now responsible only for starting and stopping the mixing cycle every twenty minutes and for adding 'sundry' ingredients which were difficult to add automatically. However, the job no longer required traditional baking skills and the doughmen lost their status as skilled 'master bakers' and were re-graded as 'mixer operators'.

The mixer operators received only intermittent feedback on the performance of the mixing process and were able to exercise little control over it. Even though this stage was critical to the rest of the production process, the operators appeared to have difficulty visualising the consequences of their actions, or lack of them, for later production stages. Neither could operators trace and diagnose equipment faults, for which skilled technicians were necessary. As a result repairs took longer to perform. The operators' experience of work was repetitive and boring and they became apathetic and careless. This was exacerbated by the fact that they were unable to develop knowledge or skills that would have made them promotable (Buchanan and Boddy 1983: 189).

In this instance the outcome of management decisions on job content and work organisation was to create a 'distanced' relationship between the worker and the technology, characterised by deskilling and a generally alienating experience of work. Buchanan and Boddy regard this kind of outcome as characteristic of 'nearly automated' systems where residual tasks, which engineers have as yet been unable to automate, are allocated to a human operator (1983: 189).

It might have been possible in the case of the doughmen to mitigate the worst of the deskilling effects of the new technology through some form of job enlargement, for example by introducing rotation between mixer operating and the supervision of the process from the control room. In other words, managers could have chosen to offset the worst effects of a technology which was designed to replace the need for human intervention by designing jobs so that they incorporated a broader range of tasks. Boddy and Buchanan refer to the content of jobs of this sort as 'limited distancing' (1986: 97).

A contrasting outcome was evident at the baking stage of the biscuit-making process. This case illustrates how the capabilities of new computer-based systems can complement human problem-solving skills. Here a computerised check-weighing machine was introduced into the wrapping and packing stage of biscuit production. The technology replaced the need to check the weight of biscuit packs manually, but also generated new information which could be used in the control of work operations. For example, the weight of biscuits was determined by a number of variables at earlier stages in the production process. Where packs were too heavy or too light, alterations had to be made by line-operators to avoid packets of biscuits being rejected. This relied on information being communicated back down the line. Under the manual system this took twenty minutes and the information received did not indicate the extent of any problems. The computerised system provided a far more rapid and complete feedback of information (Buchanan and Boddy 1983: 177–82).

This was particularly significant for the job of the 'ovensman' who baked the biscuits. Information from the wrapping and packing stage on weights now took five minutes to feed back to the baking stage. It was now possible for the ovensman to initiate rapid adjustments to the oven controls or to the machine cutting dough into the required biscuit shape and thereby to reduce the number of rejected biscuit packs. In this case, rather than replace the need for human intervention, the new technology provided information which complemented the skills and abilities of the worker in controlling the production operation. The ovensman's job in this respect became more of a supervisory role responsible for the operation of the equipment and the baking process (1983: 192–5).

Wilkinson (1983: 41–54) also reports an example of management decisions which resulted in the 'limited distancing' effect noted above. In this case, which involved an optical company manufacturing spectacles, a computer had been introduced to set automatic machinery used to cut lenses. Previously, setting had been a manual task accomplished by a skilled worker who would use trial-and-error methods to produce an accurate lens. The computer eliminated this guesswork, produced accurate lenses in one cut and reduced scrap rates. Other automatic machinery had also been introduced which again eliminated the need for informed interventions by skilled workers.

However, whilst keen to purchase the latest equipment as it appeared on the market, key members of the management team were not intent, as was apparently so in the case of the mixer operators at United Biscuits, on using the new technology in an attempt to reduce their reliance on informed human intervention. As Wilkinson points out:

... one might have expected the managers to attempt to take advantage of the *control* possibilities. That is, the quality and quantity of production could have been controlled directly by management, rather than relying on the judgement, skill and discretion of the workforce. In particular, skills could easily have been concentrated in the hands of a very small elite of technicians, and operators transformed into unskilled, cheaper and replaceable labour.

(1983: 45, original emphasis)

In fact, key managers were keen to minimise the deskilling effect of the new technology. In particular the general manager of the factory had a strong commitment to employee participation and involvement. As a result a system of job rotation was introduced so that workers moved from job-to-job on a regular basis. This had positive advantages of minimising the amount of contact individuals had with the tasks which required little skill, while creating a more multi-skilled and flexible workforce. Motivation and involvement in the work were increased and the possibility of errors due to lack of attention was reduced.

However, in another example Wilkinson provides an illustration of how a management intention to use new technology as a means of reducing its reliance on human labour can have counter-productive consequences. This case concerned a plating company. Plating involves dunking metal components into a variety of solutions to plate them in zinc, chromium, aluminium and so on. Prior to the introduction of new technology a plating line required three workers to load, control and unload the machinery. They operated levers which controlled the movements of jigs and barrels holding components as they were dunked in various vats of solution. Managers argued that the workers were lazy and unreliable, taking too much time off work, spending too long on breaks, working slowly, and sometimes leaving components in the vats for too long. All this led managers to the conclusion that tight supervision of the platers was required.

New electronic control devices, which enabled the movement of components in and out of vats to be controlled automatically rather than manually, were therefore seen by managers as an ideal opportunity to reduce their reliance on human intervention in the plating process. The automated lines required only two workers to load and unload, and the quality and pace of their work was now regulated by the control equipment. For good measure the electronic control panels were deliberately housed in a room at the back of the factory to prevent the platers having access to them! However, during the debugging of the equipment at the implementation stage of change it was frequently necessary for the platers to use the manual override on the automatic system. Once the automated lines were fully operational, platers still continued with this practice, arguing that manual intervention was required because they could get a better finish on products and reduce the scrap rate – something which they regarded with some pride.

Management's attitude however was that the platers were unreliable and 'bolshie' and that they should normally simply 'stand and watch' the process, intervening if things went wrong. Wilkinson quotes the works manager's views which aptly capture the assumptions which lay behind the management objectives in using the technology:

Some operators still do not adjust to the automatic lines. They use the manual override if you don't watch them ... It's a bad habit. It's difficult to get it into their heads that automatic is the best way. They think they can do a better job manually.

(1985: 442–3)

The outcome in this case was similar to that reported by Buchanan and Boddy in the case of the mixer operators. Again, the form of work design around the technology had been intended to reduce the need for the informed intervention of unreliable humans. The parts of the process requiring decision-making and discretion was automated and the remaining manual tasks were left for human operators. In the platers' case these involved the loading and unloading of machinery; in the mixer operators' case the starting and stopping of the process.

One view of events like this, as Wilkinson suggests, is that as less labour was required the new technology had made the process more efficient by increasing productivity. The loss of job satisfaction and control on the part of the remaining workers was an unfortunate side-effect. However, in the platers' case, argues Wilkinson, if a different set of managerial assumptions had prevailed, the technology could have been used differently. For example, the control equipment could have been located at the machine or more effort made to automate the mundane loading and unloading tasks, leaving the platers with control over the process itself (1985: 433). This might have been more consistent with the kind of improvements in product quality that managers were seeking. However the route that was taken, in response to the platers' attempts to retain control, was to continue with even tighter supervision of the work.

Computer technology in the office

In Chapter 1 it was noted that the office is a major area for the application of computing and information technology. For writers who have adopted a labour process perspective, office automation represents the attempt by management to subject clerical and other office workers to the kind of Taylorist regime that shop-floor workers had to endure in earlier phases of industrial automation. Some, albeit rhetorical, support for this view is provided by the often-quoted declaration of Franco de Benedetti, managing director of Olivetti, the multi-national office technology company. In his view, 'information technology is basically a technology of co-ordination and control of the labour force, the white collar workers, which Taylorian organisation does not cover' (quoted from Gill 1985: 42). Many writers have also suggested that managers' intentions in this respect are facilitated by the concentration of women in low-graded routine office jobs (see for example Downing 1982; Barker and Downing 1985). However, as we argued in Chapter 3, the extent of Taylorist influence on management decisions over work design has frequently been overstated. How far then is this opinion on the way work should be designed around new technology in the office borne out by the actions of managers in practice?

One of the first studies of the introduction of word processing technology in an office environment was conducted by Buchanan and Boddy, once again as part of their Scottish case studies. This particular study concerned the introduction of word-processing equipment by Y-ARD, a marine engineering consultancy (Buchanan and Boddy 1983: 145–72). Y-ARD produced technical reports for clients. These reports were long and complex, were written by several authors and often went through several revisions. In order to improve the productivity of typists and reduce the number of typing staff a word-processing system was installed.

The management working party responsible for implementing the technology debated the merits of a centralised versus decentralised form of work organisation.

On the one hand managers wanted to be able to control the flow of work to and from the word processors, and a centralised form of organisation with all the word-processing terminals pooled together made this easier. A disadvantage was that additional supervisory staff would be required. However, this solution was preferred to the option of decentralising the terminal locations so that each group of authors had their own word processor and typist. Decentralisation would have meant, among other things, better contact between the report authors and the word-processor operators and a greater familiarity on the part of authors with word-processing facilities. However, these advantages were outweighed in the eyes of the working party by several disadvantages, in particular, the lack of management control over the operation of the system (1983: 152–6).

As a result, it was decided to group the operators into two word-processing centres. Whereas previously the operators had worked with conventional typewriters for groups of report authors, they were now under the control of the word-processing centre supervisors. As noted in the previous chapter, the skills and expertise required of operators using word processors can involve new kinds of knowledge and understandings compared with conventional typewriters. However, the effect on job content of the decision to centralise work organisation was that, although the operators had greater discretion over the quality of their output, they enjoyed less control over the scheduling of their work, and the range of tasks they performed was both narrower and more fragmented. Far more time was now spent typing, and reports were not followed through from start to finish by individual operators. The outcome of these arrangements was that operator productivity was improved, with output six times higher compared with conventional typing; fewer staff were required; and management control over workflow was increased (1983: 166–71).

However, looked at in a different way, the centralisation of work organisation undermined the effective use of the technology because the word-processor operators were now both geographically and organisationally separate from the authors of the reports they were typing. In consequence the operators were less likely to be familiar with the handwriting and styles of the authors. When errors were spotted in copy by the operators they would 'have a go' at making the correction, knowing that if they got it wrong it could be changed, which was easier than trying to find the author in a large building.

From the authors' point of view this meant draft reports required more corrections, and turn-around time on report typing was not perceived to be quicker. As a result dissatisfaction increased with the way work had to be progressed through the word-processing centres and the authors took to 'bending the system' in a number of ways. For example, some authors bypassed the system and had urgent work typed by conventional means. Others had copy typed by conventional means before sending it to the word-processing centre in order to reduce errors! Finally, rather than send reports back to be altered on the word processor, corrections were made by hand. These were meant to be reported to the word-processing centre so that the master file on the computer could be updated, but this was sometimes overlooked. This meant that customers received reports slightly different from the ones stored in the computer – a potentially dangerous procedure where safety factors were involved (1983: 164–5).

In sum, the problems involved in using the word-processing system at Y-ARD revolved around the loss of contact between the typists and the report authors. This resulted directly from the implications for job content of management choices over work organisation. At first sight management's decisions in this respect might appear to bear some affinity with the views of Franco de Benedetti quoted above. However, it is important to note that management's concern to increase control was aimed at the operation of the system itself and not exclusively at the co-ordination and control of labour. Even if management's concerns in this respect had been more focused on labour control, Buchanan and Boddy provide no evidence to suggest that in general management practices at Y-ARD were influenced extensively by Taylorism or its derivatives. In fact the effects of managers' decisions on job content in this case are described in terms of what Buchanan and Boddy refer to generically as 'frustrated complementarity'. That is, in seeking to increase operational control by centralising work organisation, the choices made by managers resulted in job definitions which prevented the most effective use of the capabilities of the technology by the word processor operators.

Potentially the technology could have enhanced and extended the skills of the operators and increased their ability to meet the requirements of the authors (1983: 247). As an alternative managers could have adopted a decentralised form of organisation and allocated operators to author groups. This would have enabled far greater author–operator contact and alleviated some of the problems which had been created by the centralised system of work organisation. However, the adoption of this alternative approach would have made it more difficult to control work flow and productivity. Four years after the system was introduced the operators were still grouped in supervised pools (Buchanan 1986).

The preference for centralised forms of work organisation was promoted originally by IBM as the best way to make use of its word processing systems. The influence of this philosophy has been noted in other studies of the use of word processors and electronic data-processing (EDP) equipment. In the case of word processors it has been argued that centralisation not only enables increases in operator productivity, but also eliminates what are regarded as the non-productive and expensive personal relations and contacts between the operators, usually female, and the authors of documents, usually men (for a discussion of this issue see Downing 1980; 1982; Barker and Downing 1985; Webster 1986).

In this connection Hazel Downing reports on the use of word processors in a San Francisco branch of a large US bank. Her account reveals how, when carried to extremes, the effects of centralised forms of work organisation when used in pursuit of labour control objectives may be even more counter productive than when such forms are used in pursuit of operational control objectives. In this case a central word processing centre had been established and work stations were laid out in rows with about two and a half square feet of space between each of the 110 mainly female word processing operators. However, to supplement this pattern of work organisation a rigorous system of labour control was also developed. Operators' movements in and out of the room were strictly controlled by supervisors, and maximum productivity was sought by specifying set times for toilet allowances and bans on talking, eating and drinking at the work stations. The bank management combined this rigid discipline on the movements and actions of word-processor operators with a systematic monitoring of their productivity through devices in the equipment which counted keystrokes. Operators were paid according to their productivity, and the standard rates had been raised over a five-year period from 5000 to 12000 keystrokes per hour.

Not surprisingly this intensification of the effort bargain had increased tension among the operators as they pushed themselves to increase output. However, senior managers soon became worried when productivity started to decline. Downing's discussions with the operators revealed a considerable amount of dissatisfaction with the content of their jobs:

> The operators had grievances, and when these grievances weren't dealt with they developed their own ways of showing their dissatisfaction. They discovered that strong magnets can erase computer tape and even though it meant they had to do their work again, they derived some interim satisfaction from knowing that the bank's deadlines weren't being met and the customers weren't satisfied. And occasionally, when things got bad, the word would go down the line and 110 operators would all stand up at the same time and go to the toilet. But when things became really unbearable, they would 'forget' to key-in their personal reference code which was necessary for their productivity count, and would proceed to key-in obscenities which would come out on the print out to the customer, and because the work was standardised, no one knew who had done it.

> (1982: 50)

It is no wonder, perhaps, that Downing discovered senior managers were rethinking their approach to the organisation of work around their word-processing system, although the impression she gives is that their desired solutions would have involved no change to job content but even tighter supervision of the operators' work.

Further evidence of the effect on job content of management decisions to centralise work organisation is provided by Crompton and Jones who examined EDP applications in a bank, an insurance company and a local government office (1984: 42–77). The bank utilised a high level of on-line computer systems. The insurance company and the local government office used mainly batch processing systems, the former having a far higher level of computerisation than the latter. Crompton and Jones suggest that extreme task specialisation is facilitated by EDP technology in so far as clerical tasks are broken down into their constituent elements to facilitate the coding, punching and processing of data in preparation for input to the computer in batches. The need for accuracy and speed to meet deadlines and the cost of computer time mean that clerical staff have to work to tightly specified rules and routines which are built into the system itself [1] This was illustrated at the insurance company where work tasks were highly fragmented. As a result, clerks had little understanding of the insurance business and little or no conception of how their work related to the overall operation of the organisation or even their particular department.

However, the variations in the extent of task specialisation in different sections of the local government office showed that management decisions over work design were a crucial influence on the extent to which the content of jobs comprised only specialised tasks. In the salaries section the jobs of clerks involved a wide task range as a result of decisions made by the Assistant County Treasurer, while in the Social

Services department the responsible manager organised work so that tasks were fragmented. The bank utilised more advanced on-line interactive computer technology, which in principle allowed clerks more control over data in so far as they could both directly input and access information. However, under the direction of overall bank policy a high degree of functional specialisation between different grades of clerks remained.

CAD systems in engineering drawing offices

It was noted in the previous chapter that, in similar fashion to clerical work, the adoption of CAD systems in engineering drawing offices has been viewed by some writers as an opportunity for management to extend the Taylorist deskilling philosophy to white-collar technical workers. Indeed, the authors of a survey of firms who have adopted CAD in Scotland argue that their evidence suggests that drawing office management has adopted CAD with deskilling firmly in mind. This study found that, following the adoption of CAD, drawing office staff had less autonomy, their work was more machine-paced, they had to work faster, and the length of the working day was maximised by the introduction of shiftworking (see Baldry and Connolly 1986; also Arnold and Senker 1982).

However, different findings were in evidence in research on the introduction of CAD by one of the present authors. This study showed that the effects of decisions on job content and work organisation are likely to be more varied than the 'Taylorisation' thesis predicts (see McLoughlin 1986; 1987; 1988). The four organisations studied (see chapter 4) included a builder, an engineering consultant, a mechanical engineering firm and a ship yard. The forms of work organisation found varied along two dimensions. First, whether the systems were operated by 'dedicated' or 'non-dedicated' operators, and second whether the CAD work stations were 'centralised' or 'decentralised' (see Figure 6.1).

The approach chosen by the CAD manager in the building firm was to create a centralised CAD department to service regionally based architects and to replace a conventional drawing office at the firm's national headquarters. In fact the form of work organisation had initially been an area of non-decision. The first work stations were purchased and left in the corner of the drawing office in the hope that staff might become curious enough to use them! However, a CAD manager was then recruited explicitly to devise an alternative to this situation. He decided on a dedicated operator approach in which individuals were responsible for producing, correcting and modifying CAD drawings from rough sketches made by the regional architects prior to their issue to the construction site.

The job content of drawing-office staff was essentially confined to that of a 'CAD operator'. There was little scope for individual discretion and a narrow task range. Jobs had a low level of technical complexity and mainly involved the production of repetitive drawings of a high volume and highly standardised product. In order to maximise system utilisation and minimise the turn-around time on drawings, a twenty-four hour, three-shift system was imposed. A questionnaire survey of the nine drawing-office staff indicated a considerable discontent with the lack of opportunity provided by their jobs to engage in 'detail design' tasks and with the over-emphasis placed on repetitive drawing work – although the flexible pattern of working enabled by the shift system did find some favour among staff whose leisure interests included golf and sailing!

Figure 6.1   Forms of work organisation used with CAD

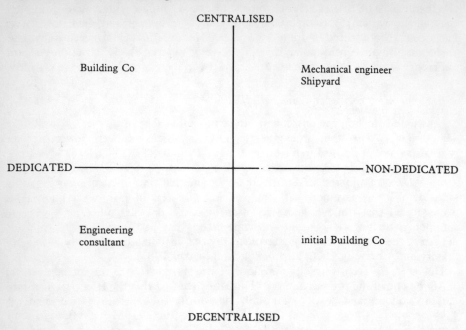

Note: 'Centralised' refers to situations where all workstations are grouped together either in separate offices, or in a screened-off area of an existing office. 'Decentralised' refers to a situation where workstations are located amongst existing manual drawing boards.

Source: McLoughlin (1986)

Managers in the engineering consultants also opted for a dedicated operator approach. However, in this case CAD workstations were located among conventional drawing boards in the existing drawing office. This was a deliberate choice by management to avoid creating a distinctive group of CAD operators which might form the basis for individual requests for more pay. Again, the jobs of CAD operators exhibited a narrow task range involving mainly routine drawing tasks. The jobs involved the drawing of rough sketches supplied by design engineers and then subsequent alteration and modification involving little scope for individual discretion. However, because the work stations were located in the main drawing office there was less discontent on the part of the operators who were able to move around the office much as they had done in the past. In addition, the CAD operators retained access to a conventional drawing board. Managers, however, were concerned about productivity being lower than expected and were considering the introduction of a shift system to increase system utilisation.

The mechanical engineering firm and shipyard adopted different approaches again. In both cases a centralised CAD centre was installed and conventional drawing-office facilities retained. However in each firm management, in response to union pressure to maintain access to new skills for all its members, adopted a non-dedicated operator approach. This involved the retraining of all drawing-office staff, who could then book time to use the CAD centre facilities. Decisions as to whether and when to use the CAD system or conventional drawing boards for a job were usually taken in consultation with drawing-office supervisors and project managers. In practice this created a problem for the CAD centre managers who could not directly control when the system would be used. This meant that system utilisation was lower than was needed to satisfy financial management's investment criteria, which demanded that high-cost capital equipment be used to the maximum. To this end both firms were considering a move to shiftworking.

Managers in both cases cited technical and operational reasons for adopting a centralised installation, in particular the need to provide a 'controlled' environment in terms of ergonomic and lighting conditions. However, it was also clear that centralisation was seen as essential to avoid drawing-office supervisors exercising complete control over system utilisation. In fact in both these cases, and particularly in the shipyard, indications were given of a conflict between the role of CAD centre managers and some drawing–office supervisors. On the one hand the CAD manager believed that the supervisors were not competent to decide whether their staff should use the CAD system or not and suspected that they discouraged them from doing so. On the other hand, drawing-office supervisors were suspicious that the CAD manager was usurping their role when staff were using the CAD equipment.

This conflict between CAD management and drawing-office supervisors appeared to be a significant factor undermining the effective use of the technology. This was especially the case in the shipyard which had installed a CAD system with modelling capabilities. To be used most effectively this system required users to develop composite skills which eroded traditional boundaries between design, drafting and manufacturing and cut across conventional divisions between different engineering disciplines (see Chapter 5). The development of these skills and abilities was constrained by the form of work organisation adopted which meant that the content of individuals' jobs was defined in terms of existing functional boundaries and disciplinary specialisms. This issue is discussed further in the next chapter.

Summary

The case-study findings discussed above illustrate that, whilst the characteristics and capabilities of technology have an independent influence on task and skill requirements, the content of jobs and pattern of work organisation are also subject to managerial choice in the design of work. Where managers have been found to consider questions of job content and work organisation, their decisions appear to have been motivated in some instances by a desire to use new technology to increase the predictability and consistency of the operation of the production process and to reduce their reliance on human intervention. We have argued that the pursuit of such objectives cannot, as many labour process writers suggest, be explained exclusively in terms of a commitment to control labour *per se*. Rather, such intentions can reflect broader operational control objectives concerned to improve the performance of the production process as a whole.

The implications of the pursuit of control objectives appears to vary, in part, according to the characteristics and capabilities of the technology concerned. For example, in instances where production systems are 'nearly automated', the introduction of new technology can facilitate the replacement of human intervention by automating work tasks and imposing constraints on the extent to which any resultant 'distancing' effects can be alleviated by alternative forms of work design. The use of new technology to replace human skills and abilities in this way was illustrated in the case of the computerised recipe desk and the doughmen at United Biscuits, and the EDP batch processing systems and the clerks in insurance, local government and banking. Attempts to alleviate the distancing effect through designing work in order to enlarge or enrich jobs were evidenced by the cases of the skilled craft workers and the computer-controlled machines in the optical company, and the clerks and the EDP system in the salaries department of the local government office.

Where technologies have new information and control capabilities they offer far more opportunities and scope for management to design jobs and work organisation in ways which use new technology to complement rather than replace human skills and abilities. Evidence of managers choosing to use technology in this way was, however, apparent only in the case of the ovensman and the computer checkweigher in United Biscuits. The cases of the word processor operators at Y-ARD and in the San Francisco bank, and the various forms of work design adopted by the four firms using CAD, all illustrate attempts to use the technology in ways which, to varying extents, led to 'frustrated complementarity'. In these cases it can be suggested that an overemphasis on the pursuit of control objectives undermined the effective use of the technology.

Finally, it was noted in the case of the building firm introducing CAD that in some circumstances managers may not make decisions at all, despite being confronted by choices over how to design work around a new technology. Indeed, similar cases of 'non-decision' have been reported in other studies, suggesting that it is dangerous to assume that questions of job content and work organisation are necessarily uppermost in the minds of either senior or lower-level managers when new technology is introduced (see for example, Storey 1986; Gourlay 1987; Batstone *et al.* 1987).

Trade union and workforce influence over job content and work organisation

So far the role of management choice in the design of work around new technology has been examined. However, it would be wrong to see the design of work as purely a reflection of management intentions. The processual model of technological change adopted in this book suggests that outcomes are not only chosen but can also be 'negotiated' in the widest sense of the word. Management intentions can therefore be modified or changed, either by the influence of trade unions or the informal influence of workgroups and individual workers. The key point here, as Bryn Jones points out, is that 'management cannot construct, *de novo*, the conditions under which labour is to function' (1982: 199). Indeed, existing circumstances may act as a considerable influence on the nature and actual extent of change that occurs.

Trade union and worker influence over job content and work organisation may be exerted in at least the following ways (see Batstone *et al.* 1987: 57): first, by influencing management choices over strategic issues such as the type of equipment or system to be introduced; second, by influencing employment issues concerned with the grading of jobs and the occupational division of labour resulting from such gradings; third, by directly influencing the content, organisation and control of work, whether by formal or informal means. The conclusions reached in Chapter 4 suggest strongly that union influence over strategic and control issues has not generally been significant. However, where existing collective bargaining arrangements and the sophistication of trade union organisation have allowed, unions have been able to exert some influence over employment issues. This section will explore these conclusions in more detail, focusing in particular on the role of the workforce itself in the negotiation of job content and work organisation on a day-to-day basis.

The significance of informal workforce negotiation over job content and work organisation has been most extensively documented in a number of case studies of the use of NC and CNC machine tools in small-batch engineering (see Jones 1985; Wilkinson 1983; Buchanan 1985a; Batstone *et al.* 1987). Developments in the use of computer-controlled machine tools began in the 1950s with the introduction of numerically controlled (NC) machine tools in the cutting of metal work pieces (see Chapter 1). Significantly, Braverman (1974: 197-206) argued that the development of NC machines was a classic illustration of the design and use of technology to deskill work along Taylorist lines (see also Noble 1979; 1984). In his view conventional machine tools involved skilled craft workers in both the 'conception' of a task (for example the interpretation of design drawings) and its actual execution (for example the manual control of the cutting actions of the machine). NC machine tools, however, store the necessary information to perform the cutting actions automatically on a punched tape installed in the control cabinet of the machine.

Whilst it is perfectly feasible for the tasks of programming and operating to be combined in the machinist's job, Braverman claimed that capitalist imperatives to deskill and degrade work meant that jobs were fragmented into their constituent components of programming and machine operating. The task of programming was allocated to 'parts programmers' located in planning offices who required no knowledge of machining operations. The job of the machinist was merely to set up the machine and push the required buttons to start and stop its operation. This resulted in the destruction of the craft knowledge of the machinist by separating the task of conception (programming) from execution (operating).

One of the first studies in Britain which examined the use of NC machines – and the more-advanced CNC machines using electronic controls – was conducted in the late 1970s at five firms located in Wales and the south-west of England (see Jones 1982). The firms supplied parts and equipment for the aerospace industry. In contrast to Braverman, the changes in job content observed in the study by Bryn Jones were indicative of a redistribution rather than destruction of skill. Moreover, the pattern of redistribution was varied amongst other things by the informal negotiation of working practices and work organisation between parts-programmers (organised by TASS) and machine operators (organised by the AUEW).

The first thing that Jones discovered was that, in practice, the skills required to program and operate NC machines were not as fragmented as Braverman claimed. In particular, programming tasks required a prior knowledge of metal-cutting tools and materials to be worked. In many cases this could be gained only through shop floor experience, while machine operating tasks required knowledge of tools, for example to detect from their sound when in operation if they were worn and required replacement. Secondly, during the routine operation of the equipment, the programming process had two points at which informal interventions by workers could be made that allowed them to 'claw-back' skills. These points were: 'prove-out', when the program written in the planning office was tested on the machine; and 'editing', where modification to programs was required.

Both these points provided opportunities for machinists to exercise discretion. For example, by enforcing union-sanctioned job demarcations they could insist that only machinists and not the programmers should activate the machine's controls during prove-out. Similarly, on night shifts when programmers were not available, machinists could use the manual override to operate the machines conventionally. Significantly, Jones observed that in firms using CNC machines there may be also potential for machinists to seek to claw-back informally skills by further encroaching into the grey areas of prove-out and editing during parts programming.

The phenomenon of 'claw-back' was studied in more depth by Barry Wilkinson (1983: 55–71) at the plant of a machine-tools manufacturer in the West Midlands. The firm had introduced nine CNC machines of various types. As Wilkinson noted, the principal difference between CNC and NC machines is that the control cabinet is equipped with facilities for the prove-out and editing of tapes. These controls provide far more flexibility in programming by allowing the rectification of faults to be accomplished at the machine on the shop floor. In contrast, prove-out and editing of NC tapes requires frequent trips back to the planning office and the storage of corrections in the machine's own computer memory. Moreover, one of the nine machines in this case was also equipped with a prototype 'manual data input' (MDI) control system. This allowed not only prove-out and editing to be carried out at the machine but also the initial programming itself. In effect the need for a planning office to carry out programming is eliminated with the MDI–CNC type of machine.

When introducing the technology neither senior managers nor management–union agreements provided detailed guidelines on job content and work organisation, although the rationale justifying investment in CNC advanced by production engineers had been based on reducing the need for skilled operators and taking control away from the shop floor. The principal union involved – the AUEW – had been concerned primarily with the allocation and grading of jobs using the new machines, insisting that only 'skilled' craft machinists operated the CNC machines. However, it had not made any attempt to influence whether the tasks themselves should involve craft skills. As a result considerable room for manoeuvre was left for the parts-programmers and machinists to resolve informally issues surrounding the precise content of their jobs and the division between them.

In this case a conflict for control over the prove-out and editing of CNC tapes occurred. Symbolically the conflict revolved around the keys which gave access to the control cabinets on the machines. The programmers claimed that the machines had been designed for separate programming and operating and that prove-out and editing, as well as any subsequent adjustments due to variations in raw materials and so on, were tasks in their domain. Therefore the keys should be under their control to stop operators tampering with the programs. The operators argued that the machine design allowed for operator input and that as skilled machinists they had the greatest knowledge and most shop-floor experience to make adjustments at the machine. Therefore, the keys should be left at the machines at all times so that adjustments could be made by the machine operators. In practice, with the tacit agreement of middle management and supervisors, the precise division of tasks was subject to *ad hoc* negotiation and re-negotiation, with the machine operators usually maintaining the upper hand. However, such conflicts were not evident in the case of the MDI–CNC machine. Here the absence of any technical constraints preventing programming being carried out on the shop floor meant that, after an intensive two-week period of instruction, the operator was able to manage the machine on his own initiative.

A study of the use of CNC machines in a Scottish engineering plant reported by Buchanan (1985a) revealed a similar finding. Here turners claimed that the new machines could offer them scope for using their craft skills, but that the form of work design adopted by management, involving a strict separation of programming tasks from machine operation, constrained this possibility. However, it was possible for the turners to make minor alterations to the programs and in this case the control cabinets on the machines were not kept locked. This enabled the turners to sustain their craft knowledge of machining techniques and combine this with an understanding of programming methods. Thus, rather than job content being deskilled, the turners were able to retain if not add to their expertise on an informal basis. Indeed, according to Buchanan, these interventions were essential in order for the technology to be used effectively. Programmers often decided to use inappropriate methods or made mistakes in preparing the program, while workpieces could be of non-standard dimensions or the machine tool could read the program incorrectly. Without a willing and able operator on hand, these errors and unforeseen contingencies could lead to dangerous and expensive results[2].

What is most significant about this case-study evidence in the context of the conclusions reached earlier is the absence of any formal trade union influence over job content and work organisation. In fact, the role of the unions in the above cases should not be surprising. In Chapter 4 it was seen that, despite in many instances being a stated intention of national policy, negotiation over the control issues highlighted by the introduction of new technology had not been sought widely by unions at workplace level. Rather it appeared that they had normally focused on the employment issues of job security, pay and grading.

However, the weaknesses of this approach to bargaining are clearly demonstrated by these instances of the introduction of CNC. For example, the retention of craft skills by the machine operator seems to be a key factor in the safe and efficient operation of the CNC machines and, it would appear, is an essential prerequisite for operator programming, especially on MDI–CNC tools. If in the above cases the AUEW and TASS had been concerned about the content of jobs as well as the demarcation between them, then the need for craft skills to be maintained in this way might have been a significant negotiating point. In the event, the overriding concern with employment issues meant that a significant bargaining opportunity – and one that might have contributed to the more effective use of the technology – was lost. This poses the question as to whether different outcomes would have resulted if a new technology agreement had been negotiated.

The potential advantages that could have been achieved in the case of CNC machine tools are illustrated by the findings of Batstone *et al.* in their case study of a small batch engineering firm (see 1987: 117–55). The firm concerned had introduced nine machining centres and three MDI–CNC machine tools into its spares plant during the early 1980s. One of the machining centres was also equipped with full MDI facilities. Unusually, note Batstone and his colleagues, the AUEW shop stewards representing the craft machinists in this case proposed and, after twenty months of discussion with management, successfully negotiated a new technology agreement. This included procedural clauses committing the management to joint consultation over company plans to introduce new technology (including the choice of equipment), a commitment to change working practices only after mutual agreement, and an agreement that only skilled AUEW members should, not only operate, but also program the machines. In the event, since the new technology agreement was concluded only after many of the new machines had been installed, the union had little influence over the planning of change or the actual choice of the new machines.

However, the agreement on operator programming served to crystallise managers' thinking on this issue, although they were already largely persuaded of the view that the operation of the technology required the use of craft skills and that, in the absence of an existing programming department, programming should be done on the shop floor. In contrast to the findings of the studies summarised above, therefore, the AUEWs attempt to defend craft status rested on the preservation of craft control over the content of the job itself, as well as on the issue of how jobs were to be allocated. That they were successful in this case was clearly due in part to the negotiation of a new technology agreement which helped confirm management's acceptance of the viability and merits of operator programming.

The defence of status and skill demarcations through the control of grading and occupational access is indicative of the approach taken historically by many craft and ex-craft unions faced with technologies which alter the task and skill requirements of their members' jobs (see for example, Turner 1962; Penn 1982; Lee 1982). The fact that the effectiveness of this traditional approach is open to question in the face of contemporary technological change can be illustrated further by events in the British newspaper industry. In fact, the union representing the major craft groups in the composing area within the industry, the NGA, has provided a classic example of the defence of craft status in the face of accelerating technological changes which threatened to transform not only the tasks which comprised the content of compositors' jobs, but also the need for a separate composition stage in the production process altogether[3].

The changes to task and skill requirements generated by new 'cold type' computing technologies in the composing area of newspaper production have already been described in Chaper 5. It will be recalled that these replaced the manual tasks and skills required by 'hot metal' methods of production[4]. A key element of the NGA's response to new technology has been aimed at protecting its monopoly of keyboard work – whether this work involves using a Linotype keyboard or keyboards linked to computers. This has been sought by seeking acceptance of the principle of the 'second keystroke' which in effect recognises the sole right of NGA members to input copy for composition even though the new technology requires skills which are possessed or can be acquired by non-compositors – i.e. they are standard keyboard skills used already by journalists and (predominantly low-paid female) typists. The NGA has sought to establish this principle by preventing members of other unions, or non-union labour, being deployed in this way. In the short term this has meant seeking bilateral agreements with other printing unions that this work was the monopoly of NGA members and in the long term seeking to form a single union for the printing industry as a whole (see Gennard and Dunn 1983; Dunn and Gennard 1984; Martin 1981; 1984; Gennard 1987). By the end of the 1970s this strategy had resulted in the achievement of agreements with a number of national newspapers, most noticeably following the fourteen-month 'lock-out' at Times Newspapers Ltd, which resulted in a recognition of the NGA's monopoly and the principle of the 'second keystroke' (see Routledge 1979; Martin 1981).

As Lee (1982: 160) points out, the NGA is an exceptional case of a craft group which has managed to socially construct a skilled status in the face of technological changes which have progressively eroded the requirement for traditional craft skills. However, as he and others have observed, where it is not based on any demonstrable or equivalent technical skill requirements, such a status, and the methods of occupational control (in this case a pre-entry closed shop) used to sustain it, are highly precarious and fragile (see also Dunn and Gennard 1984: 25–39). Indeed, the effectiveness of the NGA's strategy has crumbled during the 1980s in the face of new competitive pressures, the availability of more advanced technology and more hostile labour laws designed in part to outlaw the closed-shop practices of the print unions (see Dunn 1985). In addition there has been a greater willingness of newspaper proprietors such as Rupert Murdoch and Eddie Shah, to abandon attempts at negotiating change (see Dickinson 1984; Melvern 1986). Even though most national newspapers have sought a negotiated agreement on the introduction of new technology, the achievements of these two proprietors appear to have brought to an end craft resistance to new technology in the industry.

Survey evidence on changes in job content

This chapter has deliberately examined the issue of changes in job content by referring initially to case-study evidence. The reasons for this lie in the complex and multi-dimensional nature of the changes in tasks and jobs when new technology is introduced. As Batstone and Gourlay note, whilst surveys provide a much broader and generalisable picture than case studies, by their nature they reflect only the knowledge and perspectives of the respondents questioned. This knowledge may be partial, distinctive and difficult to corroborate. Responses on complex issues such as job content, and in particular changes in skill requirements, need to be treated with a good deal of caution, because questions of what is meant by skill and how changes in skill are measured, are left entirely to the respondent to resolve (Batstone and Gourlay 1986: 225). It should be borne in mind then that, although comprehensive in scope, the findings of surveys on this topic are particularly difficult to interpret meaningfully.

The WIRS survey asked managerial and shop steward respondents, but not the workers concerned, a range of questions on whether technological change had increased, decreased, or had no effect on job content as measured by changes in job interest, level of skill, range of tasks, and level of responsibility. Changes in worker autonomy and control were also elicited and these findings are discussed in Chapter 7. The general picture provided by the survey in relation to job content (see Tables 6.1 and 6.2) was that, where change had occurred, the skill content of work had

Table 6.1   Managers' accounts of the impact of new technology on the jobs of manual workers

Column percentages

	Job interest	Skill	Range of activity	Respon-sibility	Pace of work	How they do their jobs	Super-vision
More	46	42	38	33	22	21	16
No change	40	42	46	54	48	47	64
Less	11	15	15	12	28	31	19
Not stated	3	1	1	1	2	1	1
Change score	+70	+50	+50	+40	−10	−20	+10[a]

Base: establishments with 25 or more manual workers and experiencing advanced
technical change

Unweighted	*458*	*458*	*458*	*458*	*458*	*458*	*458*
Weighted	*212*	*212*	*212*	*212*	*212*	*212*	*212*

[a] A positive score on this item represents *less* supervision
Source: Daniel (1987)

Table 6.2   Managers' accounts of the impact of new technology on the jobs of office workers

Column percentages[a]

	Job interest	Skill	Range of activity	Respon-sibility	Pace of work	How they do their jobs	Super-vision
More	60	55	59	39	34	32	10
No change	27	39	29	55	45	38	70
Less	5	2	8	2	16	26	17
Not stated	8	4	4	4	5	4	3
Change score	+120	+110	+110	+80	+40	+10	+10[b]

Base: establishments with 25 or more non-manual workers and experiencing advanced
technical change in the office

Unweighted	*977*	*977*	*977*	*977*	*977*	*977*	*977*
Weighted	*500*	*500*	*500*	*500*	*500*	*500*	*500*

[a] See note C
[b] See note to Table VII.I
Source: Daniel (1987)

been increased, in particular for office workers (Daniel 1987: 275). In the case of manual workplaces, the majority of managers reported that new technology had had no impact on any of the nominated aspects of change in job content (see Daniel 1987: 152–61). Indeed, in most cases managers reported that change had not occurred in nearly half of the job content and control items listed.

However, where change had occurred, managers' accounts suggested that the introduction of new technology had resulted in positive effects on job content as measured by changes in job interest, skill levels, task range and responsibility. This is reflected in Table 6.1 by the high positive change scores associated with these items. These views on the positive effects of new computing and information technology were also reflected in shop stewards' accounts. In the case of office workers, managers were far less likely to report that new technology had had no impact, and far more likely to report positive effects on job content and to suggest that office work had been enriched (see Daniel 1987: 151–66). This is illustrated in Table 6.2 by the far higher positive change scores than those for manual workers in Table 6.1.

In general, managers' accounts were supported by those of office shop stewards. However the differences in their responses were also revealing. These are important because they give some clue to the role being played by respondents' perceptions in resolving the definitional problems noted above. In the case of manual stewards they were far more likely to say that there had been increases in skill levels, task range and especially responsibility. However, they were less likely to say that job interest had increased. Managers of manual workplaces on the other hand tended to be clear that job interest had increased but less likely to say that skill levels, task range and responsibility had increased. A similar pattern of variation between managers' and stewards' accounts was evident in the case of offices – stewards again tending to emphasise those positive changes in job content that might be associated with claims for regrading and more pay, and managers tending to play down these and instead emphasise the improved job interest that office workers now enjoyed (Daniel 1987: 166).

The reason for the difference, according to Daniel, is likely to lie in the negotiating positions adopted by management and unions over the issue of job regrading when new technology is introduced. On the one hand managers might be expected to emphasise the benefits of new technology to workers in terms of intrinsic benefits such as job interest, but to play down its effects on aspects of job content – skill, task range and responsibility – which might be the basis of claims for higher pay and grading by unions. Equally, it might be expected that shop stewards would place their emphasis on the latter changes and be less likely to draw attention to any intrinsic rewards to workers in terms of increases in job interest (Daniel 1987: 160–1).

This variation in managers' and shop stewards' accounts suggests that an important element in respondents' answers to surveys could be their concern to attribute a differing 'skilled status' to jobs. Their accounts cannot therefore be seen as providing an objective measure of changes in job content, although they may be a reflection of such changes. This points to the problem of comparison between survey and case-study evidence on this question. On the one hand surveys may tap the tendency for managers and stewards to define skill in terms of the status (indicated by pay and grading) attached to jobs whilst on the other hand case studies, by examining the actual changes in tasks and skills in particular circumstances, can reveal far more about the actual changes in the content of jobs that have occurred.

These findings can be contrasted with the contemporaneous, though less extensive and not strictly comparable, survey of shop stewards by Batstone and Gourlay (neither managers nor workers were questioned). Their questioning covered changes in skill, effort and worker control resulting from the introduction of new technology (see Batstone and Gourlay 1986: 225–34). Changes in terms of skill and effort are of most concern here, and the changes reported in terms of 'worker control' are again discussed in Chapter 7. The general impression from Batstone and Gourlay's evidence (see Table 6.3) is that stewards were far more likely to report that change had occurred as a result of the introduction of new technology than the stewards and managers sampled by the WIRS. The findings suggest that in most instances skill had increased as a result of change and that this was most likely to have been the case for non-manual workers, especially in maintenance. The overall pattern in the case of effort levels was for these to have increased in both non-manual/public sector and manual/private sectors. Moreover, these increases in effort were related strongly to increases in skills and, as will be discussed further in Chapter 7, worker control.

Batstone and Gourlay's survey allows further light to be thrown on the role of managers and unions in negotiating the outcomes of changes in job content. For example, they suggest that where increases in skill occurred these are explained, at least in part, by the strength and strategies of trade union organisation. Moreover, they suggest that the tendency for increases in skill and control to be associated with increases in effort levels indicates that managements are not pursuing strategies to extract greater effort by deskilling and exercising greater control over workers. Rather, it is argued, 'they are using a "strategy" – consciously or otherwise – which is the direct opposite of this. This may be part of the bargain implicitly or explicitly worked out between management and unions: that is, the content of jobs is improved, and in exchange workers achieve higher effort standards' (1986: 232).

Conclusion

This chapter has examined evidence on the way changes in job content and work organisation have been shaped by social choice and negotiation during the process of change. The evidence discussed in Chapter 3 suggested that many decisions concerning job content and work organisation fall in practice to line managers concerned with implementing and using new technology. Several of the cases outlined in this chapter indicate that where this occurs, there has been a tendency for those responsible to aim at reducing reliance on informed human intervention. While this tendency may take advantage of the capabilities of new technology to replace certain manual routine task and skill requirements, it may also ignore new task and skill requirements which enable the technology to complement human problem-solving and decision-making abilities. Management choices can undermine the effective use of new technology where they attempt to improve the performance of production operations by reducing reliance on skilled and informed human intervention. As Buchanan and Boddy have observed:

> Computing technologies make demands on human information-processing and decision-making skills, reduce the need for some manual effort and skill, and introduce new forms of work discipline and pacing. The extent to which responsibility, discretion, challenge and control increase, however, depends on managerial decisions about the organisation of work. Control objectives ... may interfere with operators' ability to use the technology effectively.

> (1983: 246)

Table 6.3 Percentage of establishments where new technology has brought about a change in job content

	Skill			Worker control			Effort			Health and safety standards		
	In-crease	No change	Fall	In-crease	No change	Fall	In-crease	No change	Fall	In-crease	No change	Fall
(a) Non-manual/public-sector group												
Finance	54	26	20	33	28	39	26	54	20	18	49	33
CPSA civil service	55	34	11	16	57	27	26	63	11	12	58	30
SCPS civil service	55	36	10	23	53	25	33	61	7	3	66	31
CPSA Telecom	43	37	20	20	34	46	71	20	9	14	40	46
POEU Telecom	61	13	26	11	58	31	40	43	17	23	69	8
(b) Production group												
Print	40	27	33	42	32	26	41	44	14	44	37	19
Chemicals	74	23	4	48	32	20	51	35	14	48	40	12
Food and drink	66	34	–	30	46	24	49	22	30	56	39	5
Engineering	64	31	5	39	41	20	39	42	19	32	58	10
(c) Maintenance group												
Chemicals	77	14	10	26	64	10	47	49	4	42	54	4
Food and drink	88	11	2	39	52	9	51	44	5	37	57	6
Engineering	74	17	10	31	57	12	49	42	10	29	67	5
Electrical engineering	73	15	12	44	41	16	59	31	9	42	49	9

Source: Batstone and Gourlay (1986)

However, the evidence also suggests that the absence of a detailed consideration of job content and work organisation in management strategies could lead to outcomes which allow workers to exercise over time a considerable influence over job content. In some instances this facilitated the informal claw-back of skills by workers and workgroups who were able to re-negotiate the content of their jobs so that the technology could be operated in a way which complemented rather than replaced their skills and abilities. The case-study evidence also supports the conclusion suggested in Chapter 4 that trade unions – or at least craft and ex-craft unions – have not sought in general to negotiate over the control issues surrounding questions of job content and work organisation. Rather, they have attempted to protect their members by seeking to restrict access to new jobs to existing job-holders. While this may have preserved or even increased the skilled status of members' jobs in terms of pay and grading, it does not guarantee that the content of jobs remains skilled. Indeed, it could be argued that the more such attempts to defend skilled status result in outcomes that do not correspond with the actual task and skill requirements of the new jobs, the more precarious a strategy this appears to be in the longer term.

Notes

1 A detailed examination of the skill content of clerical work (defined in terms of the degree of control over the work involved and the extent to which supervision or control over other organisational resources was involved) was also carried out. Crompton and Jones' findings revealed that ninety-one per cent of clerical grades interviewed did not exercise any control in respect of their own work, and eighty-four per cent did not exercise control over other workers or organisational resources. 'These jobs require', the researchers argue, 'only the capacity to read and write, and the ability to follow instructions.'

   Moreover, the extent of this apparent deskilling was related to the level of computerisation in the three organisations studied. In the local government office where the extent of computerisation was lowest, twenty per cent of clerks exercised control over their own work. In the most advanced and extensively computerised organisation, the bank, none of the clerks exercised control in this way. In summarising their findings the researchers argue that the vast majority of clerical jobs required 'little in the way of skill; work tasks were on the whole governed by explicit rules and few could exercise discretion or self-control in their work' (1984: 64).

2 This point is also recognised by David Noble in his study of NC machine use in the United States. He poses the question, 'What will a machine operator, "skilled" or "unskilled", do when he sees a $250,000 dollar milling machine heading for a smash-up?' The answer, suggests Noble, is either 'he could rush to the machine and press the panic button, retracting the workpiece from the cutter or shutting the whole thing down' or 'he could remain seated and think to himself, "Oh look, no work tomorrow".' (Noble 1979; quoted by Boddy and Buchanan 1986: 95).

3 See Chapter 5, note 3.

4 Cockburn claims that composition under 'hot metal' methods required a high degree of craft skill. For example, the ability to operate ninety keys laid out on a keyboard quite different from the standard 'QWERTY' typewriter, as well as discretion in the hyphenation and justification of lines of type and minor maintenance of the Linotype machines (Cockburn 1983: 48–53). However, Martin argues that the level of skill required under traditional methods was in fact more limited than this implies. Although speed in using the Linotype machines was important this was not, he suggests, a precondition of

recruitment. Moreover, the task itself was often less complex than in other branches of printing. This suggests that, even under 'hot metal' methods of production, it is arguable if the defence of compositor's 'skills' by the NGA rested to any significant extent on the actual skill content of the compositor's job.

CHAPTER 7

New technology, supervision and control

Chapters 5 and 6 have examined changes in work tasks, skills, job content and work organisation when new computing and information technologies have been introduced. This chapter concludes the discussion by exploring the implications of new technology for the way work is supervised and controlled. Clearly, if the introduction of new technology changes the content of work this has implications for the supervisory management. Woodward, for example, predicted that advances in technology would be associated with the development of impersonal and integrated 'mechanical control' systems which would reduce the need for the personal supervision of work operations. Similarly, many labour process writers associate the deskilling of job content with an increasing centralisation of management control and the replacement of the need for direct supervision of the labour process by automatic monitoring devices built into the equipment. Other writers, such as Blauner (1964), have suggested that advanced automation is associated with increasing work group autonomy and a decentralisation of management decision-making.

This chapter starts by exploring the nature of the management choices that are opened up by new technologies in relation to the control and supervision of work. Evidence drawn from a range of recent studies of the adoption of new technology is then discussed. This evidence shows once again the importance of choice and negotiation in shaping particular outcomes, also that workers themselves can have a decisive influence at critical junctures during the process of change.

Management choice and the supervision and control of work

In Chapter 1 the principal capabilities of new technology as a means of capturing, storing, manipulating and distributing information were outlined. As Child notes, given such enhanced information handling capabilities, 'it would be difficult to overstate its relevance for the processes of integration and control within organisations' (1984: 257). What then are the choices that managements have in relation to the supervision and control of work when new computing and information technologies are introduced? The choices involved fall into three related categories: management organisation – in particular the extent to which operational control is centralised or delegated; the role of supervision – in particular the extent to which workers are to be subject to direct personal supervision or impersonal 'mechanical controls' incorporated within the technology; and work organisation – in particular the extent to which decisions are devolved to the work group allowing them to become 'self-supervising' or autonomous.

Management organisation

When large main-frame computers were first introduced into work organisations in the 1950s it was predicted that this would lead to an increase in the control over production operations that could be exercised by senior management, who would tend to centralise decision-making and reverse trends towards delegating decisions (see for example Leavitt and Whisler 1958). However, it has also been suggested that computer technologies, in particular on-line systems, would lead to the decentralisation of control closer to the point at which work operations are carried out, and thus to a decrease in senior management control (see Robey 1977 for discussion). One problem in the debate, as Child points out, is that 'delegation' is often confused with 'decentralisation'. The latter suggests a devolution of senior management control to subordinates. Delegation, on the other hand, implies a specific form of decentralisation where 'authority to make specified decisions is passed down to units and people at lower levels in the organisation's hierarchy' but control over decision-making is still retained by senior managers (Child 1984: 146).

In other words, trends towards centralisation and delegation of decision-making responsibility are both consistent with the increase in central management control enabled by computing technologies. It is also important to recognise that centralisation and delegation are not simple dichotomies (1984: 146). Thus, when introducing new information technologies, choices about management organisation refer in effect to the degree of emphasis given to centralisation and delegation in the allocation of decision-making authority.

In practice a number of arguments for and against emphasising centralisation and delegation can be made, whose salience will vary according to organisational circumstances. According to Child (see 1984: 146–53; 260–1) for example, whilst centralisation offers some efficiency gains in terms of senior management's desire to oversee work operations, it may lead to 'overload' at the top of the organisation. Delegation, although leaving more scope for independent action by subordinates, offers significant advantages in terms of the flexibility with which lower-level managers can respond to circumstances and their motivation in exercising responsibilities.

Allowing for different organisational circumstances Child suggests that computing and information technologies can open up a range of new possibilities for delegating decisions. A similar point is made by Boddy and Buchanan, who point out that senior managers can use new technology to enhance the confidence of their own decision-making, taking advantage of the greater visibility of subordinates' actions to monitor closely and even intervene in the decision-making of lower-level managers. Taken to extremes this can actually have the opposite effect on their subordinates, placing more pressure on them and making the technology appear as a threat. On the other hand, where the new technology is used to support the delegation of decisions it can also offer improvements in visibility and confidence to lower-level managers and supervisors as well (1986: 158–61; also Buchanan and McCalman 1988). Finally, it should be noted that associated with the centralisation and delegation of decision-making may be choices in relation to the simplification of organisation structure, and in particular a reduction in middle levels of management and clerical staffs who were previously required for routine information-processing tasks but whose work can now be accomplished automatically (Child 1984: 262; Boddy and Buchanan 1986: 158).

Supervision

The role of supervisors has been a longstanding problem in British industry, reflected in supervisors' declining status and their increasingly ambiguous relationship to management. A study published in the early 1980s concluded that failure on the part of British management to deal with the problem of supervision had resulted in them becoming 'lost managers' because, although their range of responsibilities had increased, the authority invested in them by management to execute these tasks had declined (Child and Partridge 1982). As already noted, some writers have argued that the information and monitoring capabilities of new computing and information technologies enable management to substitute 'impersonal', 'mechanical' (Woodward 1980) or 'technical' control (Edwards 1979) devices for the more problematic personal controls provided by direct supervision. Other writers have argued that such 'mechanical controls' may become 'mixed in' with existing supervisory tasks (Dawson 1987a; Kinnie and Arthurs 1988). Dawson argues that management choice is not simply a question of whether or not to replace first-line supervisors with mechanical controls, but in fact involves redesigning 'composite first-line supervisory tasks' around 'a mixture of traditional practices and new computer-oriented activities' (1987a: 58).

A further issue is the degree to which there should be what Child (1984) terms a 'high degree of supervisory emphasis' involving tight personal supervision of work operations, or whether workers should be given more autonomy. Friedman (1977) argues that management choices on the degree of supervision may be contingent upon product market conditions and the labour market position of the particular workgroup concerned. However, one point that will emerge from the case study evidence reviewed below is that 'team autonomy' needs to be distinguished from what Friedman calls 'responsible autonomy'. The former is an outcome which is the result of management strategy, while the latter may be an outcome of both the independent technical constraints on work organisation and the informal influence of work groups as well as management strategy.

Implicit in these choices is the question of the training supervisors should receive and the role that they should play in the introduction of new technology. Should they, for example, be given the same training as their subordinates in order that they fully understand the work of those they will be supervising, and as part of this process be fully involved in the actual management of change during implementation? Or should supervisors be given an appreciation of the changes taking place, but during implementation remain responsible for current operations and not become sidetracked by the process of change itself?

Work organisation and work groups

It is important not to overstate the capabilities of the 'mechanical controls' provided by computing and information technologies. For example, as Gallie points out, the idea of 'mechanical control', at least in automated process settings, is often taken to imply that because information is provided to workers through dials and the like, this necessarily governs their actions. However, such information normally simply tells workers what action to take, it does not ensure that they actually carry it out (Gallie 1978: 214–16)[1].

In fact, Gallie finds more persuasive the rather different argument about the effects of advanced automation technology on the supervision and control of work suggested by a contemporary of Woodward, Robert Blauner. Blauner (1964) put forward the proposition that advanced automation technologies lead to many of the traditional control functions of supervision being taken over by the work group. In his view, as with Woodward, the need for direct personal supervision becomes redundant. However this occurs, not because the mechanisms of control are built into the technology itself, but because the system interdependencies between work tasks created by advanced technologies require more collective collaboration between workers. Because the quality of work performance depends on the collective efforts of the work group rather than the individual, control is exercised through what amounts to 'self-supervision' by the work team itself (see Gallie 1978: 220).

Gallie's own research in highly automated oil refineries led him to conclude that automation may be 'conducive to a certain degree of team autonomy' (Gallie 1978: 220). However, he was at pains to argue that team autonomy was not purely, or even mainly, a product of the capabilities and characteristics of automated technology. Rather, the managements concerned had a choice as to whether or not to reduce the amount of direct personal supervision and to allow the development of more autonomy among the work team. In the cases he studied, managers decided that direct supervision was increasingly unnecessary as the training and experience of the work teams improved, and that cuts in supervisory staff would allow them to produce the savings needed to reduce costs as the industry became more competitive (Gallie 1978: 221).

Summary

To summarise so far, managers have a number of choices in relation to the supervision and control of work operations. First, there is a choice between the degree of centralisation and delegation of management control. Secondly, there are choices in relation to the role of the supervisor – in particular the extent to which work is subject to close personal supervision and the extent to which 'mechanical control' devices are substituted for supervisors. Third, there are choices about the extent to which decisions are devolved to the work group. Of course, these choices may be interlinked. For example, a decision to centralise control may also be linked to a decision to increase direct supervision. However, centralisation may also be linked to policies to substitute impersonal 'mechanical' or 'technical' control devices for direct personal supervision. Similarly, delegation may involve more decision-making authority being delegated to either supervisors, work groups or both.

Finally, it is important to emphasise once more that whatever choices managers may be faced with by the introduction of new technology, it does not follow that their eventual actions will be part of a coherent strategy. As will be seen below, management approaches to the question of how work is supervised and controlled tend, as in the case of questions of work design, to be issues which are delegated to lower levels of line management. As a result, rather than being a core element of corporate thinking behind the decision to introduce new technology, it appears that issues such as the role of supervision are dealt with, if they are not ignored altogether, only during or even after the technology has been implemented. As in the case of work design, this can leave considerable scope for individual workers and work groups to influence outcomes at critical junctures during the process of change.

Changes in supervision and control

This section draws on recent case-study research which has examined changes in the supervision and control of work when new computing and information technologies have been introduced. The first two cases discussed are drawn from work in which the authors have been involved with colleagues in the New Technology Research Group (see McLoughlin *et al.* 1983; Dawson 1986; Dawson and McLoughlin 1986; Clark *et al.* 1988; Dawson 1987a; 1987b). The remaining examples come from case studies and surveys conducted by other researchers.

The first case provides an example of a management decision to take advantage of the increased visibility of work operations and confidence in decision-making to delegate day-to-day operating decisions closer to the point of operations. In this case the need for a new second-line supervisory role to replace the existing hierarchical management reporting structure was identified. However, whilst the new supervisory role enabled delegation to take place, the threat of management and trade union resistance prevented any simplification of the organisation structure.

The second case provides an example of a management which did not anticipate the implications of technological change for the role of the first-line supervisor. In this instance the technology concerned was highly conducive to the development of more flexible forms of work organisation based around an autonomous work team. In the absence of any coherent managerial or union pressures to the contrary this further eroded the role of the first-line supervisors and allowed the workgroup to establish informally new forms of work organisation. However, subsequent changes in senior management thinking recognised the possibilities for redesigning supervisory roles under more advanced systems.

Computerisation and the redesign of supervision in railway freight

The first example involves the introduction of the computer-based freight information system (TOPS) by British Rail, which began during the early 1970s and has already been referred to in Chapter 3. It will be recalled that the management sub-strategy developed to implement the technology was based around a multi-disciplinary 'task force' to manage the project's implementation. This enabled a particularly tight 'coupling' to be achieved between corporate objectives and the manner in which change was implemented at workplace level. However, this case also illustrates how, even where management has a relatively coherent strategy for implementing new technology, this can still involve major omissions in management thinking over how a technology is to be used. As will be seen below, the role of the 'task force' during implementation and its subsequent disbandment had a direct bearing on the outcomes of subsequent attempts to make changes in management organisation and the role of supervision.

'TOPS' (Total Operations Processing System) provides an excellent example of the potential of computing and information technologies to make work operations more visible to management and to improve the confidence of management decision-making (see McLoughlin *et al.* 1983). The system had been introduced in an attempt to combat operational control problems which had resulted in the highly inefficient use of locomotives and wagons. Prior to computerisation, information on the whereabouts of wagons and locomotives was collected by making daily physical checks in marshalling yards and freight terminals. This information was then reported by supervisors to one of around twenty divisional control offices who received similar reports from other yards and terminals under their control. This information was collated in order to provide a 'picture' of the current distribution of rolling stock in the division, and then reported to one of five regional control offices who constructed a similar picture for the region as a whole. Finally, at national headquarters a picture was then put together from regional reports of the distribution of rolling stock over the network as a whole.

The information provided by this manual reporting system was used by managers at divisional, regional and headquarters levels to plan the day-to-day movement of rolling stock and to allocate empty wagons to customers. However, there were two major problems with this system. First the transmission of information, which took place by telephone and telex, was very slow. In the context of 'time-sensitive' railway operations where trains are constantly moving, this was a highly unreliable and inefficient way of controlling the distribution of rolling stock. Second, the accuracy of the information received by management was highly suspect. For example, because the manual reporting system was so inefficient, local supervisors would 'hide' wagons in remote sidings to ensure that local requirements could be satisfied. The presence of this stock was not reported to senior managers and so contributed to a vast over-provision of wagons on the network given the amount of freight actually carried[2].

The introduction of a computerised freight information system offered the potential means for transforming this situation. The TOPS system was based around a main-frame computer at national headquarters. This was linked to on-line data terminals located in around 150 reporting centres situated in marshalling yards and at freight terminals. Each wagon and locomotive was given a computer code number and every time a train left or arrived at a location, or a wagon was loaded or unloaded, this information was communicated for input to the computer to the local reporting centre by staff using short-wave radio or a facsimile machine. This enabled the central computer to maintain a real-time picture of freight operations throughout the network. The architecture of the software was designed to maintain accuracy by cross-checking reports so that any attempts to input bogus information would be rejected.

In addition some aspects of operational control were automated. For example, by programming timetable information into the computer, automatic prompts could be given to staff at local level to prepare booked trains for departure. Similarly the daily task of allocating empty wagons could also be accomplished automatically. Local operations were thus made far more visible to senior managers who now had more rapid and accurate up-to-the-minute information available at their finger tips simply by making an enquiry at a computer terminal. As one senior freight operations manager commented, 'we now had a production line we could control'! How then did managers set about achieving increased operational control in practice?

While the 'task force' had developed a particularly effective sub-strategy in implementing the technology, it was only towards the latter end of the implementation stage that the project management team started to consider the operational control issues that would be involved in using the technology once the system became fully on-line. In other words, in the initial phase of implementation managers were primarily concerned with the technical task of installation and, commendably in this case, the retraining of staff to work with the system. These concerns tended to override the issue of how management itself would need to change, for example in terms of the role of supervisors, although an early assurance had been given to the rail unions that the system would not be used to control labour by monitoring the whereabouts and productivity of drivers and guards. Thus, in support of the arguments already put in Chapter 3, management's strategy towards the control of work operations was not fully conceived at the time of the decision to introduce the technology at the beginning of the process of change, nor was this strategy driven by a need to increase control over labour. Rather a sub-strategy concerned with changing management organisation *emerged* during the process of change and was focused on improving the overall control and performance of freight operations.

The principal problem in this respect was how far should day-to-day decisions and the planning of local freight operations be delegated. That some degree of delegation rather than centralisation was required was apparently beyond argument. Any attempt to use the new computer system to centralise freight operations control at headquarters or even regional level would impose massive overload problems for senior managers in a national organisation the size of BR. In fact, the belief of railway operators within BR had traditionally been that, ideally, day-to-day operating decisions were best taken at the point where events occurred by the person 'on the spot'. However, in practice the need to gather together information on what was happening in a given area and use this to co-ordinate operations – especially where a major operating contingency such as a derailment occurred – meant that prior to computerisation the divisional level was the most appropriate to which most day-to-day decisions could be delegated.

Computerisation opened up the possibility of a further delegation of responsibility (Dawson and McLoughlin 1986; Dawson 1986; 1987b). One option was to broaden the responsibilities of the traditional first-line yard supervisor to cover the operations in the hinterland of the marshalling yard. Authority for day-to-day operating decisions covering this hinterland could then be delegated to them. The opportunity for further delegation had become apparent during the initial stages of the implementation programme. Senior managers in charge of the 'task force' noticed large piles of computer printout containing information on current freight operations accumulating in the corner of the office while visiting a reporting centre. They knew this information could be used to inform local decision-making and plan the movement of local freight trains. However, nobody had formal responsibility for this task which would involve making continuous enquiries on the computer terminals in the reporting centres. It was therefore decided to create a completely new second-line supervisory post of 'area freight assistant' (see Figure 7.1).

The project management team's image of the area freight assistants' role was of a 'new breed' of supervisors who would act as local 'supremos'. In practice area freight assistants were formally in charge of freight operations in a geographical area around the reporting centres which usually included at least one marshalling yard. They worked from a desk in the reporting centres on which was located a computer terminal. They used the terminal to acquire information on operations in the area, for example on such things as loaded wagons awaiting departure, empty wagons that were surplus to local requirements, and the services scheduled for departure and those approaching. This information was then used to pre-plan local movements of traffic in accordance with marshalling operations, themselves pre-planned (by the yard supervisor – see below) to meet the schedule of incoming services.

This new supervisory role required a radical change in the orientation of supervisory decision-making. Instead of being parochial and concerned with dealing with crises as they occurred, area freight assistants needed a much broader

Figure 7.1   Levels of supervision before and after computerisation

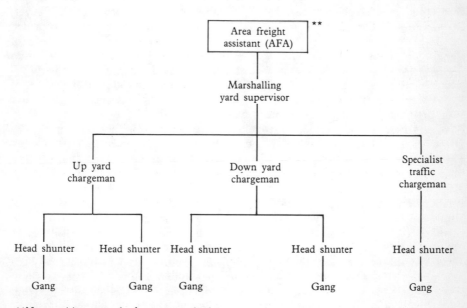

**New position created *after* computerisation

Source: Dawson and McLoughlin (1986)

perception of local operating requirements and the consequences of their actions for the movement of freight around the rest of the network. At the core of this role was the idea that supervision was now more impersonal and centred around the use of 'real-time' information from the computer system to direct operations. Ideally the project management team saw this as a role more suited to management grades – possibly a job appropriate as the first post for graduates finishing BR's graduate training programme. In practice, however, union pressures and other expedients meant that the post was given a supervisory grade, albeit a higher-level one, and that a mix of staff were recruited. These included staff who had worked in divisional and regional control rooms, ex-members of the implementation team who installed the technology and trained the staff – many of whom were graduate entrants – and some yard supervisors.

Computerisation also had implications for the existing supervisory roles in marshalling yards (see Dawson and McLoughlin 1986; Dawson 1987a; 1987b). Overall responsibility for the control of marshalling-yard operations was vested in the yard supervisor, a first-line position located in the yards. They were assisted by deputies ('chargehands') responsible for sub-units of the yard and 'head shunters' in charge of the shunting gangs in the yard itself (see Figure 7.1). The role of the yard supervisors was to some extent eroded in that they were now responsible for reporting directly to the area freight assistant and their performance in supervising marshalling-yard operations was now highly visible.

In the case of chargehands and head shunters the effect of computerisation was to incorporate knowledge they previously required into the computer system. Prior to computerisation chargehands needed detailed knowledge of railway rules and regulations when authorising train departures to ensure, for example, that dangerous goods had been marshalled in the correct train formation. However, these checks were now made automatically by the computer which provided printout on how the train should be marshalled for the safe conveyance of dangerous goods, although the chargehand still had to check that the train formation corresponded to that on the list generated by the computer. Similarly the head shunters had needed a knowledge of railway geography and routing in order to shunt wagons correctly. However, this was no longer required since the computer now provided a routing code for each wagon according to its destination.

Whilst these changes tended to erode the traditional basis of first-line supervisory authority in the marshalling yard, computerisation also contributed to more effective overall supervisory performance. The yard supervisors' and chargehands' task of pre-planning yard operations was made far easier by the availability of accurate and up-to-date information which gave their decision-making more certainty and confidence. These effects were reflected in the ambiguous attitude of some supervisors to the system. On the one hand they were conscious that they could not 'go over the computer's head', but at the same time they recognised that it enabled them to do their job more effectively. In this respect it is also important to recognise that there were large areas of the existing responsibilities of both supervisors and other yard staff which were left untouched by computerisation.

In the case of yard supervisors this was particularly so in dealing with frequent operating contingencies such as derailments and staff shortages, while chargehands retained a crucial role in leading and motivating staff. Similarly, head shunters retained control over deciding the sequence of shunting operations, as attempts to computerise this had not proved technically feasible. This meant that whilst computerisation eroded some aspects of supervisory responsibility and enabled them to execute other tasks more effectively, it also highlighted other aspects of their work as more important. Yard supervisors, in particular, saw their role far more as 'trouble shooters' resolving problems rather than dealing with routine matters. One commented that this aspect of their work was often misunderstood:

> People say you're wandering about and you're doing nothing because you haven't got a shunting pole in your hands and you're not pulling points. You can see people saying: 'he ain't got a bad job like, he walks about and don't do bugger all'. But the first sign of anything going wrong and everybody shouts out: 'where's the yard supervisor!'
>
> (Quoted from Dawson 1987a: 57)

The delegation of significant areas of decision-making from divisional control offices to area freight assistants obviously raised the issue of whether or not the old reporting system was still needed (see Dawson and McLoughlin 1986; 1988). However, while the project management team was aware of this implication when redesigning the role of local supervision, they were unwilling to challenge existing organisation structures and patterns of occupational division. On the one hand, middle managers were resistant to the idea of the abolition of divisional controls on the grounds that this would promote the creation of a two-tier management structure for all organisational functions, a long-standing objective among senior management. On the other hand, the trade unions were strongly opposed to the scrapping of divisional control offices since it would mean a loss of 1000 out of 2000 controllers' jobs. In fact railway controllers were an occupational group with strong traditions in railway operating culture, performing a task analogous with that of air traffic controllers. The project management team was unwilling to confront this problem when implementing the system, and when it was disbanded there was no central impetus from headquarters to seek further organisational changes, many believing that redundant parts of the structure would 'wither on the vine'.

In the event, it was left to each railway region to seek to accomplish further reorganisation on a piecemeal basis, a task which had not been completed some ten years after the system had been installed. As a result the new role of area freight assistant existed alongside the role of divisional controller, which to all intents and purposes it made redundant. Not surprisingly a certain amount of tension existed between these two roles, especially as area freight assistants sometimes ignored formal requirements to report certain information or clear particular actions with divisional control. Thus, rather than delegation resulting in the simplification of the organisation structure, it in fact became more complex, with elements of the old reporting structure existing alongside the newly created second-line supervisory role.

The erosion of supervision and emergence of team autonomy in telephone exchanges

Our next example returns to the study of the modernisation of telephone exchanges in British Telecom already referred to in Chapters 3 and 5. Three points arising from the earlier references to this study are of particular importance for the following discussion. First, in contrast to the BR case, senior BT managers adopted a mixture of 'top-down' and 'bottom-up' approaches to implementation which allowed considerably more scope for local managers to develop their own solutions to problems. However, this ran the risk of local managers' actions, or lack of them, leading to outcomes which were not anticipated or intended by senior managers. Second, as in the case of British Rail, the corporate strategy behind Telecom's exchange modernisation programme, although intended to improve labour productivity and allow staff reductions, was not driven purely by labour control objectives. Rather these concerns were derived from, and secondary to, operating objectives aimed at reducing costs and strategic objectives aimed at improving the range and quality of telecommunications services that could be offered to satisfy increasing customer demand. Having said this, senior managers provided little more than broad guidelines on how improvements in operational control were to be achieved in practice by local managers, and in particular what role maintenance supervisors should perform.

Third, in Chapter 5 it was shown how the system interdependencies created by the new semi-electronic TXE4 exchanges had a direct bearing on changes in the work tasks of exchange maintenance technicians. These involved the replacement of many of the manual skills associated with maintaining the electro-mechanical Strowger exchange by a qualitatively different level of mental diagnostic and systems skills associated with the maintenance of the TXE4 system. All of these factors had considerable significance for changes in the way work was organised and controlled in the new exchanges.

Traditionally, maintenance supervisors worked their way up through the ranks of maintenance technicians. As a result, since they themselves had spent many years as technicians, they had an intimate knowledge of the jobs of their subordinates. However, in reality individual technicians enjoyed considerable autonomy over their work. As long as performance targets were being achieved, day-to-day decisions over task allocation and work organisation in an exchange were frequently left by the supervisor to senior technicians. In fact, the role of the supervisors had been eroded over time by the extension of national bureaucratic guidelines which governed maintenance procedures and routines, and by the increasing use of computers to monitor exchange performance automatically. These devices, rather than the supervisor, set the parameters within which day-to-day decisions in an exchange were taken.

The principal task left to maintenance supervisors was that of administration, dealing with such things as the collation of performance information, completion of time sheets, scheduling holidays and days off, and other routine administrative tasks. In effect these 'administrative' supervisors were typical of the 'lost manager' syndrome identified above. They made infrequent interventions in the exchanges themselves and had little day-to-day control over work operations. However, the fact that they had themselves worked as maintenance technicians was an important symbol of their authority, both to them and their subordinates. Where a major technical problem did occur they could understand the nature of the difficulty and give advice, while they could also appraise and counsel apprentices and junior technicians on their progress (see Clark *et al.* 1988: 134–9).

The introduction of the new TXE4 exchanges eroded this last symbol of supervisory authority and severely exacerbated their position as 'lost managers'. This occurred for the following reasons. First, as the corporate strategy behind the introduction of the technology did not involve a detailed consideration of the role of the supervisor (something which was not challenged by the supervisors' trade union, the STE – see ibid: 64–6), there was considerable scope for local managers to develop their own solutions to the problem. In practice this meant there was a tendency to assume existing approaches would be appropriate, although in some instances attempts at developing new responses were made.

Second, the new task and skill requirements generated by the TXE4 system were manifested during the implementation and initial operation of the new exchanges by the development of a *skill superiority* of maintenance technicians over their supervisors. In order to acquire the required skills the technicians had undergone a retraining programme lasting seventeen weeks spread over a two year period, the latter months of which in most cases were spent gaining 'hands on' experience commissioning and debugging equipment in the new exchange and/or gaining experience in an already operational exchange. In contrast, supervisors typically remained in charge of the old exchange right up to the point at which the new one became operational. They were sent on an 'appreciation course' which lasted two weeks, but usually had no involvement during the implementation stage of change in the exchanges for which they were ultimately to be responsible and had little opportunity to build up anything but the most general appreciation of the technical content of their subordinates' new work. As a result they lost the technical understanding of the work of their subordinates which in the past had been the basis of their supervisory authority (see ibid: 139–46).

In some telephone areas studied these factors contributed in a number of exchanges to supervisors withdrawing even further into their administrative role. As one commented:

> TXE4 has, if anything, distanced me more from the floor. Being an old Strowger man, I've only got the sketchiest outline of the technology. I can't make a useful engineering contribution. If you've got troubles (i.e. a technical problem) you feel much happier if you know what you are talking about. I am now purely a paper engineer. On the old exchange system I definitely influenced the actions taken by the technicians. I could look at a problem and say 'take that course of action' and I knew that if it was done properly it would work. Now I can't say that.
>
> (Quoted from Clark *et al.* 1988: 144)

In a third area studied a new policy towards supervision was developed, although this turned out to be different from the approach adopted eventually at corporate level. In this case local managers took the view that the supervisor should receive the same training and hands-on experience as the technicians and specialise in the supervision of computerised exchanges (the supervisors in other areas were responsible for a variety of exchange types). As a result the supervisor was far more involved during the implementation stage of change and subsequently played a more proactive role in the supervision of exchange performance. He made regular visits to the exchanges under his control, although the technicians' skill superiority was still such that little influence was exercised over the day-to-day conduct of work operations.

In the absence of any managerial or trade union influence to the contrary, and in the context of the further erosion of the role of the supervisors, the question of how work should be organised and controlled was resolved by the informal influence of maintenance technicians in the new exchanges. The result was the emergence of flexible forms of work organisation based around team autonomy (see Clark *et al.* 1988: 160–89, for a full discussion). In part this reflected system interdependencies created by the common control architecture of the TXE4 system (see Chapter 5). These acted as a strong technical constraint on the possible forms of work organisation which could be chosen, creating a strong imperative towards more flexible and collaborative team working and self-supervision in similar fashion to the findings of Blauner and Gallie outlined earlier.

In this case, because finding faults meant being able to relate an individual symptom to the overall exchange as a total system, and because the complexity of some faults encouraged a 'two heads are better than one' approach, it was more effective for the technicians if they executed and made decisions about work tasks as a team. This situation can be contrasted to the position in the old electro-mechanical exchanges where the step-by-step architecture of the system and electro-mechanical technology used made it far easier to divide work into individual responsibilities which corresponded to geographical areas of the exchange and to which routine maintenance programmes could be allocated. The need for a systems approach and the absence of a high level of routine maintenance tasks in TXE4 exchanges meant that such pronounced division of labour, whilst possible, would have undermined the effectiveness of maintenance in the new computerised exchanges. Nevertheless, as Gallie also found, it is important to emphasise that the new forms of work organisation that emerged were the product of choice and negotiation in the context of the constraints and opportunities set by the technology – they were not inevitable outcomes in the manner implied, for example, by Blauner.

Finally, it is instructive to note senior managers' attempts towards the end of the TXE4 exchange modernisation programme to develop a new policy towards supervision. In fact two alternatives were considered. The first involved recognising a clear distinction between the technical and managerial aspects of supervisory tasks. Day-to-day technical aspects could be delegated to 'working supervisors' or 'super technicians' within the technician work group, while the managerial aspects would become the responsibility of a new first-line 'system manager' responsible for managing overall exchange performance. In effect the old first-line supervisor's role would be abolished. The other view was that the technical and managerial aspects should be merged at first-line supervisory level. This meant supervisors being fully system trained and able to direct, and where necessary carry out, day-to-day maintenance operations. In essence this meant a reinforcement of the existing role of the first-line supervisors by providing them with the technical expertise necessary to give them the authority to exercise their responsibilities.

In the event, a modified version of the 'system manager' idea was chosen, without the back-up of the 'super-technician' role which would have raised major demarcation issues with the trade unions. As preparation for this new role the technical appreciation course for supervisors was replaced by a new course on exchange management. The new course programme recognised the skill superiority of the technicians, but suggested that the authority of the supervisor rested not on technical expertise but in his or her ability to manage human resources and use performance information to direct the overall maintenance effort (see for a full discussion Clark *et al.* 1988: 190–203). In other words, rather than seeking to redesign the role of the supervisor so that they would be able to engage in direct personal supervision of the work of their subordinates, managers decided instead to develop an 'arm's length' model of supervision concerned with the overall operational control of maintenance and exchange performance. Thus, the day-to-day decisions which might have been the subject of more direct personal supervision were, in effect, formally devolved to the work group itself.

Summary

These two cases illustrate a number of tendencies which, as will be seen below, have also been identified in other research. First, there is a tendency for particular aspects of the role of the first-line supervisor to be eroded, especially those concerned with the direct control of labour. This was most evident in the BT case where their authority over staff, at least in relation to technical matters, was severely undermined by the introduction of new technology. The new skill superiority of the maintenance technicians effectively eroded the last claim to authority that the supervisor had in so far as this was based on their accumulated technical knowledge and experience as maintenance technicians. In the BR case the effects of erosion were less dramatic, although they affected a wider range of supervisory roles at and below first-line level. Yard supervisors, chargehands and head shunters all experienced aspects of the supervisory task being incorporated within the computer system and the erosion of the traditional basis of their authority.

Second, both the technologies opened up possiblities for the redesign of supervisory roles, especially in relation to the overall control of work operations. In the BR case this was clearly evidenced by the creation of a new role of area freight assistant, an office-based job designed explicitly to use information generated by the computer to control local freight operations. In the BT case the concept of a new supervisory role only emerged in senior management thinking towards the end of the modernisation programme, but nevertheless was based around the concept of a 'system manager' responsible for the overall control of maintenance operations. Related to this question was the kind of training and experience relevant to these new roles. In the BR case it was felt that the demands on the supervisor as a more computer-oriented manager were sufficiently new to call for a different kind of individual to those traditionally employed in marshalling yards. In the BT case the issue was focused on the appropriate kind of training and in particular how far the emphasis should be on managerial or technical abilities and knowledge.

Third, both cases illustrate the practical difficulties and constraints involved in making large-scale organisational changes when new technology is introduced. This was especially so in the BR example, where the possibility of removing the divisional level of the old reporting hierarchy was not pursued with any vigour, because of resistance from middle managers and the trade unions. The effect, therefore, was to ramify rather than simplify organisation structure. In the BT case the new supervisory concept implied the abolition of the first-line supervisory role and the creation of two new roles of 'super-technician' and 'system manager'. In practice, industrial relations considerations meant that only a watered-down version of the system manager role was adopted by redesigning the responsibilities of the existing first-line supervisor.

Finally, both cases illustrate the emergent nature of managerial strategies in so far as the role of supervision and the organisation and control of work was not an issue considered in detail at corporate level when deciding to adopt new technology. In both cases it was only towards the end of implementation and during the initial operation of the new systems that the existence of a supervisory problem was recognised. In the meantime, and particularly in the BT case, the position of supervisors as 'lost managers' was exacerbated.

Changes in supervision and control identified in other research

This section places the above findings in broader context by comparing and contrasting the changes in supervision and control identified with those found in other case-study research. For example, is there wider evidence of a general tendency for some aspects of the role of supervisors to be eroded and for new responsibilities to emerge? Similarly, has management used new technology to centralise or delegate operational control? Finally, is there a tendency for work groups to become self-supervising, suggesting that automated technology may be conducive to the emergence of team autonomy?

The erosion or enhancement of supervision?

Evidence on the erosion of first-line supervisory roles is provided in several of Buchanan and Boddy's Scottish case studies. This erosion occurred in three ways (1983: 249–50). First, in some instances supervisory responsibilities were incorporated within the computer system to provide the machine pacing of work. For example at Reiach and Hall, where computer-aided draughting was adopted, the management task of co-ordinating staff effort was achieved through the disciplines imposed by the CAD system itself. Similarly at Y-ARD, where word processors were adopted, prompts given by the system encouraged typists to maintain work pace. Second, in some cases production performance information was captured and analysed by the technologies. This was illustrated in the use of electronic checkweighing machines at United Biscuits which provided information feedback to operators on the production line (see Chapter 6) and in the chemical process plants owned by Ciba–Geigy where a variety of automatic recorders captured information. Third, the computer systems used widened access to information (in similar fashion to the TOPS system in BR). This meant that, whilst more information was available at all levels, performance was also more visible to managers. Fourth, in some instances managers lost their traditional skill superiority over operators (in similar fashion to the supervisors in the BT case). For example, the introduction of various computer-aided manufacturing technologies at Caterpillar's engineering plant meant in some instances that supervisors' existing knowledge was now redundant and in others that they were not able to interpret the output of the systems and were thus reliant on the knowledge and skills of other people within the organisation.

Buchanan and Boddy's case-study findings also pointed to changes in organisation structure following the introduction of new technology designed to lead to a centralisation of management control over work operations. These changes, as they were at pains to emphasise, were not determined by technical capabilities but by choices reflecting managers' underlying assumptions and values. These decisions resulted either in the creation of new specialist groups or new management hierarchies and positions. For example, planning and programming departments were established at Caterpillar and at Reiach and Hall to control and administer the use of the CAD system, whilst at Y-ARD new supervisory posts were established to control new word processing centres. The effect of this functional specialisation and structural ramification, argue Buchanan and Boddy, was to interfere with interdependencies between functions and operators' ability to use the technology effectively (see Chapter 6). The belief of line managers was that tighter control and specialisation was more effective. In only one case did management choose to simplify organisation structure. This occurred at one of the Ciba–Geigy plants where management decided as an experiment to abolish the role of the first-line supervisor. The work team was given more autonomy and overall supervisory responsibility was transferred to the plant manager and shift chemist. The success of this experiment was indicated by fifty per cent higher labour productivity than in other plants using conventional arrangements, and by a more consistent product quality (1983: 219–31).

Rothwell's case studies (see Chapter 3) provide further evidence of the neglect of supervision by managers during the introduction of new technology. However, they also reveal that in some instances supervisory roles have been enhanced (see Rothwell 1985). She found that supervisors were rarely involved in the initial stages of change when investment decisions were taken and design specifications and technologies were chosen. At the implementation stage she found that the managers responsible were more concerned with technical issues and questions relating to the staff who would use the system. Where supervisors were not the primary users they were frequently one of the last groups to which detailed consideration was given. In fact (as in the BT case) during implementation it was common for the supervisors to keep the old system going while the new system was phased in. In this respect training tended to come after that of operators or even to be ignored completely. However, where the organisation's training function was involved training tended to take place in parallel to that of operators.

The importance of the timing and content of supervisory training was illustrated by two instances. In one case – the order-processing section of a photo products firm – training was delayed and supervisors, who lacked the necessary skills, lost credibility. This was later realised to be a mistake by management. However, in another case – an engine plant – management introduced autonomous work teams on the assembly line, but in preparation for this supervisors were sent on courses which included team-building exercises. This assisted them to work out a new relationship with their subordinates which now required the supervisor to 'stand back and observe' and not intervene in team activity.

Rothwell also identified many of the facets of the erosion of existing first-line supervisory roles already noted above, for example the incorporation into the technology of supervisory tasks and the increased autonomy of work groups. However, the enhancement of supervisory roles by using new information to improve their ability to monitor performance and initiate corrective action were also in evidence. In some cases, too, the technology released supervisors from routine tasks, enabling them to concentrate more on the 'people' aspect of the job such as motivating staff, and in others they assumed the role of 'technical experts' in solving break-downs and faults with the new equipment. Rothwell also found a tendency for some simplification of organisation structure and in a number of cases two supervisory levels were merged into one.

Centralisation or delegation? The case of EPOS

Many of the issues raised in relation to the centralisation and delegation of management control were evident in research conducted by Steve Smith at the Open University on the implications of EPOS technology (see Chapter 1 for a description) in retailing (see Smith 1987b). His findings illustrate the strong influence that can be exerted by existing management thinking and organisational arrangements in shaping the outcomes of change. The research was based on interviews with store and headquarters managers in fourteen UK retailing companies with experience of using EPOS. Significantly, it was found that EPOS technology was being used in some cases to further extend the centralisation of management control over retailing operations at headquarters level and in others to support existing policies to delegate decision-making to stores.

The companies which had a highly centralised management control structure tended to be the major high-volume mass retail chains. Typically most of the major decisions relating to the performance of stores, such as the number and selection of lines carried, store layouts, promotions, prices, window displays and so on, were determined at headquarters level. Store managers were left with responsibility for such things as staff supervision and deployment, the general morale of staff, the appearance and security of the shop and its stock, and the communication to headquarters of trading information for stock control and auditing purposes. Store managers were usually regarded by headquarters management as a major problem. Their judgement in knowing what would sell best in their stores was not trusted, so senior management were keen to eliminate as much local discretion over stock ordering and buying as possible by centralising these activities at headquarters.

EPOS was perceived by headquarters managers as offering a clear opportunity to centralise further their control over store operations. Significantly, though, the potential of the technology to increase labour control, for example by monitoring the productivity of individual staff, was not seen as of paramount importance. Improved operational control could be achieved in a number of ways. For example, laser-scanners could read bar code information on products as they passed through the EPOS checkout and depleted lines could automatically be re-ordered. The automatic capture of data improved stock control and allowed headquarters management to reduce the amount of stock carried by each store. At the same time throughput was increased by concentrating on the fastest-selling lines. Indeed, prior to the introduction of EPOS the manual reporting of stock information was so unreliable that the only way management in some instances could find out what stock they had in a store was to go out on to the floor and physically count it (a situation similar to the BR case discussed above).

However, the centralisation policy also led to a number of problems. EPOS data was not made available to store managers and all buying and ordering decisions based on the new computerised information were made by headquarters. In several cases headquarters managers found themselves 'deluged with data they did not know what to do with'. Unexplained regional variations in sales patterns often occurred when the decisions of central buyers failed to result in the expected sales.

Store managers complained that this was because their local knowledge of what would sell in their stores was ignored by headquarters. In fact, the traditional 'entrepreneurial' skills of the store manager were eroded by the EPOS system, headquarters managers preferring the 'safer' marginal and incremental improvements in performance that could be achieved by centralised decision-making.

A contrasting use of EPOS was evident in another group of retailers which included one national food supermarket chain, one national men's clothing chain and four prestige retailers in London's West End. Here management control involved far more delegation of decision-making to stores, and managers in most cases had considerable autonomy in deciding such things as the numbers, types and price of lines, and the promotion and display of stocks. In these cases the 'entrepreneurial' or 'craft retailing' skills which were eroded in the other group of retailers were central to the practice of store management. In addition, the buying and selling aspects of retailing were not rigidly separated. Instead these were located at store level in the belief that knowledge about what would sell came from a close interaction with shop staff and customers. The benefits of EPOS systems in these cases were derived from the way they could support delegated decision-making at store level and assist high-quality and well-paid salespersons in the task of selling goods in a courteous and well-informed way to customers.

Smith makes the point that the decision to centralise or delegate was not determined decisively by factors such as market position, product market and geographical location, although these were important. Both approaches could be viable in similar circumstances and EPOS systems could be used to support either. However, although it is not explored, implicit in Smith's account is a suggestion that EPOS systems might have been used more effectively by the retailers who *were* prepared to delegate more operational decisions to store level. In these cases many of the problems that were experienced by centralised retailers in relation to buyers and the role of store managers were apparently absent. Again, there is a strong parallel with the BR case discussed above, where management was similarly prepared to use the capability of the new technology to make information available at operating level in order to support delegated decision-making. Arguably, if senior managers had sought to centralise the control of freight operations in the manner of some of the retailers studied by Smith, this would have resulted in many of the problems encountered by the retailers, particularly lack of flexibility at operating level and problems of 'overload' at headquarters level. Such developments in the BR case would have clearly undermined the effective use of the computer system in controlling freight operations.

The emergence of team autonomy? The case of FMS

The scope for different solutions to the problem of organising and controlling work has been explored in detail in the case of Flexible Manufacturing Systems (FMS) by Jones and Scott (1987). They compared an American and a British plant using FMS technology. This case is of interest since it points once more to the influence of existing organisational arrangements and industrial relations considerations on forms of control and supervision, but also adds further support to the view that automated systems may generate imperatives conducive to the emergence of team autonomy. In this latter respect – as in the BT case above – it provides an example of how the work group can play a key role in influencing the outcomes of change.

The American system was located in the mid-west plant of an agricultural machinery manufacturer ('Alpha'). The FMS had been in operation for over ten years. One problem initially faced by managers was that their labour contract with the trade union (as is typical in industrial plants in the USA) imposed strict rules with regard to seniority rights over transfers, redundancies and promotions. In order to get round the constraints this imposed on their ability to select staff for the new system, managers created a completely new occupational category of 'FMS qualifiers'. They then deliberately recruited staff from outside the organisation to fill these positions. One advantage of this arrangement was that it provided for continuity in the composition of the work team who stayed with the system for several years and built up a store of working knowledge. However there were also unanticipated consequences for managers' ability to control FMS work operations.

The FMS provided the flexible response needed to adjust to rapid changes in marketing or production requirements. This resulted in a number of business gains, although these were difficult to quantify. Thus, whilst establishing the worth of the system required these gains also exposed the inadequacy of conventional cost and accounting controls as a basis for evaluating the productivity and performance of the FMS operators. Second, the insularity of the FMS operators from the rest of the plant made it difficult for management to control their work directly. The FMS new skill combinations and enhanced interdependencies between machine set-up, operation, maintenance, scheduling and programming, which meant that it was essential for the operators to acquire multiple 'system level' skills embracing mechanical, electronic and software knowledge. In similar fashion to the maintenance technicians in the BT case the skill superiority established by the work team enabled them to evolve informally their own form of flexible work organisation and to 'self-supervise' their work.

These changes can be contrasted to those which were identified at the British plant ('Turnco'). This FMS system had originally been installed on the site of a machine-tool manufacturer by its parent company as a teaching and demonstration exercise. Subsequently, the FMS system was integrated within Turnco's business activities. When originally installed, the managers in charge of the FMS had encouraged the development of flexible forms of work organisation which would allow the three operators, whom they had hand-picked and put on 'staff' grades, to develop the multiple skills required. However, a decision was then made by Turnco's parent that the FMS should be absorbed into its own operations. One immediate implication was that the operators' jobs would have to be reclassified according to Turnco's existing grading structure. This was based on conventional job demarcations that accorded programming tasks with 'staff' status and jobs such as machine set-up and operation with 'works' status. Turnco's personnel management was unwilling to create a new 'FMS grade', as had happened in the case of 'Alpha'. This meant that the operators were given a choice between retaining their existing 'staff' grade or moving to 'works' grades which, although of lower status, were more highly paid because of bonuses.

In the event two operators chose the latter course. Subsequently, a far more marked division of labour emerged, in particular in relation to programming tasks which were retained as the exclusive prerogative of the 'staff' graded operator. In turn the rigidities that this division introduced into the pattern of work organisation affected the performance of the FMS, with operators no longer willing to take responsibility for the operation of the system as a whole. The situation was explained by one operator in the following terms:

> Now if I walk past the machine and I think it's been programmed to run slow I'll do one of two things; I'll either tell the programmer and he'll make the changes; or I'll just think, 'Oh, he's made a mess of that one!' and let it go rather than tell him for the sake of saving thirty seconds on the job. There's no incentive to now, really.
>
> (Quoted from Jones and Scott 1987: 34)

Jones and Scott argue that these two cases illustrate the significant influence of management policies and existing organisational and industrial relations arrangements on forms of work organisation in FMS installations. They also suggest that the kind of rigid job classifications that existed in the British case were inconsistent with the most effective use of FMS technology. However, as they also point out, the more effective flexible form of work organisation that was evident in the American case was not an outcome of conscious management choice but an informal and *de facto* arrangement decided upon by the workgroup. This in turn raises important questions for managers in developing policies for the most effective use of FMS. Where such *de facto* workgroup autonomy arises, they will have to decide whether to 'organise or enfranchise' such groups. More generally, suggest Jones and Scott, 'managers may need to surrender some power over their human factors of production in order to regain power over production operations as a whole' (1987: 36). This point is one that will be taken up in Chapter 8.

Survey evidence on changes in supervision and control

Finally in this chapter an attempt can be made to establish general trends in changes in supervision and control by examining the survey evidence from the WIRS and Batstone and Gourlay's shop steward survey. However, many of the caveats in interpreting survey evidence that applied changes in the skill content of jobs (discussed in Chapter 6) also apply here. Indeed 'control' is as difficult to define and as open to as many different interpretations as the terms 'skill' (for three contrasting views see Child 1984: 136–71; Thompson 1983: 122–52; Francis 1986: 104–30). The major problem with the measures of changes provided by surveys is that they are unable to take into account many of the complexities and different dimensions covered by the term 'control'. Again, this means that it is very difficult to interpret what the findings of surveys actually mean, or to compare them with each other.

For example, the WIRS measured changes in 'control' in terms of differences in the amount of autonomy workers had over the pace of work, how they did their job, and the level of supervision to which they were subject. In the case of manual workers, managers' accounts of changes along these dimensions suggested that on balance the introduction of new technology was unfavourable in terms of workers' control over the content and pace of work, but that they were subject to less supervision. This picture was confirmed by stewards' accounts. In the case of offices, managers reported a far more favourable effect upon the control exercised by office workers than in the case of manual workers. However, office shop stewards differed in their perceptions of the effects of new technology and reported a modest reduction in the control exercised by office workers (see Tables 6.1 – manual jobs, and 6.2 – office jobs, in Chapter 6).

In the case of manual workers the tendency for their control to be reduced was even more marked when the size of workplaces was considered, indicating that the larger the workplace the more likely that management would use the technology to reduce individual worker control (Daniel 1987: 155). Batstone and Gourlay's shop steward survey suggests that new technology has been associated with considerable change in the amount of 'control' exercised, but in contrast to the WIRS findings they found that this tended to mean an increase for manual workers in the private sector and a decrease for white-collar workers and public sector workers (Batstone and Gourlay 1986: 230).

In the light of the case-study evidence discussed above the WIRS findings at least would seem to support the view that the role of personal supervision may be eroded in so far as new technology provides an impersonal means for automatically pacing work and determining how jobs are to be executed. However, the WIRS findings do not provide any indication as to whether workers enjoyed more collective autonomy. It could be that the increased control for manual workers reported by the shop stewards questioned by Batstone and Gourlay is an indication of such an increase. The point made by Gallie noted earlier that, although new technologies may provide information which prompts workers and tells them in what order to accomplish tasks this does not necessarily mean that they will act accordingly, is of relevance here. Indeed, such apparent reductions in individual worker control may be consistent with increases in collective work group autonomy providing the 'self-supervision' necessary to ensure effective work performance. Unfortunately, the design of existing surveys on this topic does not allow any firm judgement to be made on this question.

Slightly more illuminating evidence is provided by the WIRS in terms of the measurement of changes in flexibility. Although restricted to works managers accounts in predominantly large manufacturing workplaces, these do tend to suggest that the introduction of new technology has been associated with increased flexibility in the organisation of work, mainly in the form of a relaxation of conventional job demarcations rather than the creation of new composite grades. In the context of the case-study evidence discussed above this could indicate that management-initiated moves towards fully autonomous work teams, which may reasonably be expected to involve the creation of new composite grades (such as the 'FMS qualifier' grade noted above), have not been widespread – although some of the barriers to such developments may have been eroded (Daniel 1987: 172–80).

Conclusion

The findings of the research summarised in this chapter suggest that: new technology can make work operations more visible and improve the certainty and confidence of management decision-making; it can erode aspects of existing supervisory roles; and it may be conducive to the emergence of team autonomy among work groups. In some instances line-managers have attempted to use new technology to make work operations more visible and improve the certainty and confidence of management decision-making resulting in the creation of new departments or supervisory jobs with the aim of centralising control over work operations. Where too much emphasis has been placed on centralisation there is evidence to suggest that inappropriate forms of supervision and control can prevent new technology from being used effectively.

In other instances, managers have sought to use new technology to place greater emphasis on the delegation of decision-making to points nearer the production process itself. This occurred partly in recognition of the fact that some new technologies eroded aspects of the role of first-line supervisor and were conducive to a degree of team autonomy among work groups, and partly in recognition of the possibilities that new technology provided for enhancing other aspects of supervision, in particular in relation to the overall control of operations. The evidence suggested that these types of arrangement provided a more appropriate means of controlling and supervising work operations and could lead to the more effective use of new technology.

However, a common finding was for management to have neglected the issue of how work was to be controlled and supervised when new technology was introduced, at least in the initial stages of change. In part this reflected the more general problem associated with supervision in Britain exemplified by the 'lost managers' thesis. Indeed, in some instances where management failed to come to terms with the issue of how work should be supervised and controlled it was the work group itself which was the most significant influence in developing new, and perhaps more appropriate, forms of work organisation based around self-supervising work teams. To this extent, the 'team autonomy' that emerged cannot readily be explained, as some labour process writers would argue, as the result of a management strategy designed to induce 'responsible autonomy' (Friedman 1977). Rather it provides a further illustration of the significant influence that workers can exert at critical junctures during the process of change, and of the independent influence of the capabilities and characteristics of the new computing and information technologies in enabling them to achieve this.

Notes

1 For example, in the oil refineries studied by Gallie it was found that the operators' interpretation of information from control room dials and displays was crucial in avoiding errors and anticipating problems which would have affected product quality, cost, the speed of production, and ultimately the physical preservation of the plant and the safety of the workforce itself (Gallie 1978: 212–16). Thus, whilst the technology replaced the requirement for manual intervention in the production process, it also generated computerised information about the process which the work group had to interpret in making decisions and solving problems. This made the exercise of discretion on their part more rather than less crucial since errors in interpretation could be exceedingly costly.

2 In fact twenty per cent of the half million wagons on the network went unaccounted for in the daily checks on their whereabouts. Similarly there was a twenty-eight per cent over-provision of locomotives and train crews. After the system was introduced the number of wagons was reduced by sixty-one per cent only partly as a result of a decline in freight traffic (Dawson and McLoughlin 1986).

CHAPTER 8

Conclusion: choices and alternatives in using new technology

This book has been concerned to draw together the findings of recent research on the introduction of new computing and information technologies at work. The aim has been to examine the roles played by managers, trade unions and workforce in shaping the outcomes of technological change and to highlight the day-to-day realities of these processes at workplace level. As such this book is meant as an antidote to generalised speculation about the 'social impact' of the new 'information technology revolution' which suggests that the choice facing British industry is simply 'automate or liquidate'. We would argue that the 'big bang' conception of technological change implicit in such views is of little immediate relevance to practitioners, be they managers, engineers, trade union representatives or workers, not least because it fails to reflect the complex choices and issues that arise when these technologies are introduced in the workplace.

Our focus has therefore been on micro-processes of change within organisations. We have sought to develop and apply an analytical framework which, in contrast to most recent research, also takes into account the independent influence of the characteristics and capabilities of the new technology itself. This chapter begins by summarising the principal arguments made and conclusions reached, and seeks to broaden the discussion by considering some of the alternative choices and policies that may exist for managers, unions and workforce in seeking to use new technology more effectively.

Summary of the argument and main conclusions

Chapter 1 examined the claim that recent developments in computing and information technologies constitute a new 'technological revolution', and also explored the question of whether the technology was 'new' and if so in what sense. The broad conclusion reached was that the latest developments in computing and information technologies do constitute a distinct stage in the automation of work, since they dramatically increase the possibilities for extending what has been termed 'tertiary' or 'control' automation – that is, using technology to control the transformation and transfer of raw materials.

Computing and information technologies are characterised by their ability to capture, store, manipulate and distribute information and to automate aspects of the control of work operations. Microelectronics radically increases the power, speed, reliability and flexibility but reduces the size and cost of computing and information technologies. The range of potential applications of these technologies is therefore radically increased. It can be regarded as a 'heartland' technology which can support new applications of computing and information systems in sectors of employment and aspects of the production process where the automation of control was previously not possible. In this sense microelectronics-based computing and information technologies are distinctive and new. However, these capabilities do not necessarily mean that human intervention is no longer required, rather that the control and co-ordination of work operations is likely to be *more* automated than it has been in the past.

There has been a tendency in popular commentary to view the choices facing managers, unions and workforce when confronted by microelectronics as being tightly constrained by technical and commercial imperatives. The three perspectives examined in Chapter 2 attributed rather different degrees of emphasis to the influence of choice and negotiation in shaping the outcomes of technological change. For example, Woodward's analysis of the relationship between technology ('production system'), organisation structure ('management control system') and commercial success provided little analytical space for the idea that managerial choice or negotiation with the workforce might be significant influences on the outcomes of change. Rather, the logic of technological progress and commercial requirements meant that managers were required to adapt their organisation structure to suit the production system if the organisation was to be commercially successful. Similarly the idea that technological change might have a political dimension involving a possible conflict of interest between management and labour was not seen as significant.

Labour process theory, on the other hand, emphasised the class-based conflict between capital and labour as the driving force behind technological change and identified technology as one means by which management could seek to extend its control over the shop floor. Writers such as Braverman tended to see management strategy in rather one-dimensional and mechanistic terms as involving the progressive deskilling of job content through the application of Taylorist forms of work design and control. Resistance on the part of the workforce was not seen as significant. Other contributors to the labour process debate, such as Friedman (1977), Edwards (1979), Burawoy (1979) and Thompson (1983), offered more sophisticated models, stressing the alternative choices that were available to managers in controlling the labour process and emphasising the way management strategies were shaped by the 'contest for control' between capital and labour. However, little significance was attached by labour process writers to the problems of formulating and implementing management strategies. The approach tended to assume that strategies to control the labour process flowed unproblematically from overall business strategy and that managements acted in unison to pursue a single objective.

The concept of strategic choice (Child 1972) provided an antidote to the weaknesses of both these perspectives. In this approach the forms of technology, work and management control within an organisation were seen as the result of choices made by a power-holding group, a 'dominant coalition' of managers, in accordance with their particular assumptions and values. These strategic choices could be modified by other organisational actors, particularly lower levels of management responsible for implementing decisions, and through collective action by the workforce. The idea of strategic choice derives from the social action approach in organisation theory which regards forms of work and organisation structure as emerging from political decisions made by organisational actors rather than being determined by technical, commercial or capitalist imperatives.

The concept of strategic choice and the social action approach in general have had considerable influence on recent research on the introduction of new computing and information technologies. This research has stressed the manner in which the outcomes of technological change within individual organisations are not simply the inevitable result of 'impacts' on the organisation but are shaped by the actions and choices of managers, trade union officials and workers during its introduction. A concern with the way the outcomes of change are socially chosen and negotiated during the introduction of new technology focuses attention on the *processual* nature of change. We argued that the process of introducing new technology can be broken down into a number of analytically distinct stages – initiation, decision to adopt, system selection, implementation, and routine operation – each of which provides opportunities for managers, unions and workforce to choose and negotiate outcomes.

These stages capture the temporal element of technological change. However, in any given process of change a number of procedural and substantive issues are highlighted which require decisions to be made by organisational actors, either by conscious choice and negotiation or by omission or non-decision. The point in the process of change at which the temporal stages of change intersect with particular issues raised by the introduction of new technology provide critical junctures at which managers, trade unions and workers can seek to intervene in order to influence outcomes. The remaining chapters of the book set out to explore empirical evidence on their influence over a range of issues such as changes in work tasks and skills, work design and job content, and supervision and control.

Chapter 3 examined the empirical evidence on processes of managerial decision-making when new computing and information technologies are introduced. Little support was found for the view that managers were simply acting in response to wider economic and technical pressures in deciding to adopt new technology. Similarly, the view that technological change was initiated and directed at increasing management control over the labour process was also shown to be too simplistic. Rather, managers pursued a diverse range of objectives when new technology was introduced, reflecting both their level and functional position within the organisation.

Industrial relations considerations, such as labour control, were rarely decisive in decisions to adopt new technology. Labour policies did not flow unproblematically from overall business strategies. Available evidence suggested that the introduction of new technology was not accompanied by significant innovations in labour regulation, and in general personnel specialists performed a marginal and reactive role. Senior managers' strategies acted as 'steering devices' which, although defining the parameters within which lower-level managers could act, left considerable room for manoeuvre in implementing change. In these circumstances the 'sub-strategies' developed by managers in implementing and using new technology had a critical bearing on the extent to which strategic objectives were implemented in practice.

Chapter 4 examined empirical evidence on the influence of trade unions. New computing and information technologies raised procedural issues such as when and how unions were able to influence the introduction of new technology and substantive questions concerning the issues which were or could be the subject of union concern and influence. Traditionally, trade unions have not sought to influence strategic issues in management decision-making during the initial stages of change and have been content to bargain over the terms and conditions of employment as the 'price for change' once these decisions have been taken. The implications of new technology for control issues such as job content and work organisation highlighted the shortcomings of the traditional approach. However, there was little evidence to support the view that trade unions have acted as a barrier to the introduction of new technology in Britain in the 1980s.

Since the late 1970s the TUC has advocated the negotiation of new technology agreements as a means of exerting union influence on technological change. These agreements were intended to both widen the range of issues and extend the influence of trade unions into earlier and more strategic levels of management decision-making. In practice NTAs have not been widely adopted, and even where they have been their form and content have fallen some way short of the original TUC objectives. Trade unions appear to have been able to exert most influence where the existing collective bargaining framework already enabled them to exert a strong influence over other issues – a key factor here being the degree of sophistication of union workplace organisation and the extent to which it was externally integrated with the wider official union structure. However, even in these circumstances the bargaining agenda still appeared to have been restricted to traditional employment issues such as job security and pay and grading. Where trade unions were already in a weak position, existing collective bargaining arrangements proved to be a wholly inadequate basis for the negotiation of change. Nevertheless, there was little evidence to suggest that technological change had directly weakened trade union organisation at the workplace.

The starting point for both labour process and action theory in analysing technological change was the assumption that outcomes are not determined by the 'impacts' of the capabilities and characteristics of technology. This view was in part a reaction to the analysis presented by writers such as Woodward, who had concluded that commercial success was dependent upon adapting organisation structure to suit the requirements of the technology. Labour process writers argue that the characteristics and capabilities of technology are dependent upon and explained by the objectives of capitalist management. Writers using an action theory approach argue that technology is the product of prior strategic choices. Chapter 5 went against the grain of these arguments by suggesting that an analysis of the independent influence of technology was a necessary complement to the analysis of the way outcomes were socially chosen and negotiated. It was argued that to ignore the possibility of such independent effects was to run the risk of throwing the technology baby out with the determinist bath water. Therefore an attempt was made, with the aid of two of our own detailed case studies and reference to the findings of other research, to show how the concept of *engineering system* might be deployed in the identification of the independent influence of new computing and information technologies on task and skill requirements.

It was concluded that while new computing and information technologies eliminated or reduced the range and complexity of tasks requiring manual skills and abilities, they also led to more complex tasks and new skills requiring mental problem-solving and interpretive skills – for example in using new computerised information to control work operations and in deciding upon corrective action when problems occurred. In some instances, new tasks still required tacit knowledge and abilities acquired while using older systems. Work tasks using the new electronic and computerised systems frequently appeared to involve a more abstract relationship with the technololgy than the old mechanical and electro-mechanical equipment they replaced. Thus, as Buchanan and Boddy (1983) have argued, although automating some aspects of the control of work operations, new technologies normally require a continued human presence and informed intervention if the control capabilities of the technologies are to be used effectively to complement human information-processing and decision-making skills.

The following two chapters explored the implications of choice and negotiation for job content, work organisation, supervision and control. In Chapter 6 the questions of whether new work tasks are allocated to existing jobs or grouped into new ones, and how these jobs are to be linked into a more general pattern of work organisation, were considered. A key aspect of job content is the skill required by job-holders. Choice and negotiation by managers, unions and workforce could therefore have an important bearing on the outcomes of technological change in relation to skills. The evidence examined on the role of managers revealed that many decisions concerning job content and work organisation fell in practice to line managers concerned with implementing and using new technology. Where this occurred there had been a tendency for those responsible to design work around computing and information technologies with the intention of reducing reliance on informed human intervention. While this took advantage of the capabilities of new technology to replace certain manual routine task and skill requirements, it also ignored new task and skill requirements which might have enabled the new systems to complement human problem-solving and decision-making abilities. In these cases, because the choices made by managers interfered with workers' ability to use the new technology, they undermined the effectiveness of the use being made of new systems and equipment.

However, other evidence suggested that the lack of management strategies in relation to work design could lead to outcomes which allowed workers to exercise considerable influence over job content and work organisation. In some instances this facilitated the informal claw-back of skills by workers and workgroups, who were able to renegotiate the content of their jobs over time so that the technology could be operated in a way which complemented rather than merely replaced their skills and abilities. In this respect the case study evidence supported the conclusion noted above that trade unions have not sought in general to negotiate over the control issues surrounding questions of job content and work organisation. Rather, they – or at least craft and ex-craft unions – have attempted to protect their members by attempting to preserve access to new jobs for existing job-holders. While this may have preserved or even increased the skilled status of jobs in terms of pay and grading, it has not guaranteed that the content of these jobs has remained skilled. Indeed, as illustrated in the case of the NGA, the more such attempts to defend skilled status resulted in outcomes that did not correspond with the actual task and skill requirements of the new jobs, the more precarious a strategy this appeared to be.

Chapter 7 examined changes in the way work was supervised and controlled. The evidence discussed revealed that: new technology could make work operations more visible and improve the certainty and confidence of management decision-making; it could erode aspects of existing supervisory roles; and it was conducive to the emergence of team autonomy among work groups. In some cases managers had placed too much emphasis on improving certainty and confidence by further centralisation of decision-making. This prevented workers using the technology effectively and resulted in inappropriate forms of supervision and control being retained or introduced. As a result many of the potential advantages of new technology were lost.

In other instances managers attempted to use new technology to place greater emphasis on the delegation of decision-making to points nearer the production process itself. In some cases this involved the creation of new supervisory roles concerned with the overall control of workplace operations. Where this occurred there was evidence suggested that it could lead to the more effective use of new technology. However, the most common occurrence was for managers to have neglected the issue of how work was to be controlled and supervised. In some cases this meant the work group itself had been the most significant influence in developing new forms of work organisation based around self-supervising work teams, which were more appropriate to the system interdependencies and task requirements of the new technology.

Choice and the effective use of new technology

One of the themes running through this book has been a commitment to identifying both the social *and* technical determinants of technological change within organisations. We have therefore suggested that whilst outcomes may be seen as a product of social choice and negotiation, and not as a direct reflection of the capabilities and characteristics of technology, technical factors still have to be taken into account in describing and explaining technological change. In arguing for this we are not simply suggesting that organisation structures are the product of a need to adapt to the requirements of given technologies for commercial success as implied, for example, in the work of Woodward. However, neither are we totally convinced by the view that, within the constraints of situational variables such as product and labour markets and organisation size, the choices open to organisational actors are such as to make the outcomes of change highly variable. As Littler has pointed out, the current phase of development is one:

> where the 'new' technology is genuinely new. As such it appears to be malleable and to offer a range of options – centralisation versus decentralisation; enhancement of skills versus the polarisation of skills away from the shopfloor; rigid controls versus delegation of decision-making over production.
>
> (1982: 143)

However, in our view this does not necessarily mean that all of the choices made lead to the most effective use of the technical capabilities of new technology.

Much of the research reviewed in Chapters 6 and 7 pointed strongly to the conclusion that new computing and information technologies have been used most effectively when they have complemented rather than simply replaced human involvement in the production process. In other words, whilst these technologies can replace or reduce the need for the exercise of manual effort and skill – for example where a robot is used to automate paint spraying in a car plant – many of the cases discussed confirmed that they do not necessarily eliminate the need for humans to exercise problem-solving and decision-making skills. In several instances there was still a need for skilled or informed human intervention, for example to interpret information and monitor and control work operations or to maintain automated systems and equipment. Notwithstanding this, it appeared that in many instances managers were more concerned to view the effectiveness of new technology in terms of its capacity to replace rather than complement human skills and abilities, and to improve the certainty and confidence of decision-making by relying less on the judgement and discretion of subordinates.

One explanation for the persistence of such concerns could be that they reflect a continuing commitment on the part of British managers to Taylorist ideas and practices which aim to deskill work and centralise management control. However, we saw in Chapter 3 that there is little historical evidence to support the view that Taylorist ideas have had a widespread influence on British management practice. Moreover, the concepts of management and management strategy that have been adopted in this book are rather more complex than that implied in such arguments. First, managers have been viewed, to use Wilkinson's (1983) terms, not as 'messengers' for the functional requirements of the wider economic and technical system, but as 'creative mediators' who actively shape the outcomes of change within organisations in accordance with their own values and assumptions. Second, management strategy has been seen as a processual and dynamic phenomenon. As Mintzberg (1978) observes, 'strategy' is a 'pattern in a stream of decisions' (quoted from Watson 1986: 134).

In the context of the introduction of new technology, the critical junctures in the process of change provide a 'stream' of points at which such a pattern can merge in organisational decision-making. The point we wish to make is that neither Taylorist nor neo-Taylorist ideas and practices appear to provide the overall pattern in the stream of managerial decisions which accompany technological change. Indeed, questions of labour control, and more broadly the human and organisational aspects of technological change, do not appear in the majority of instances to be the decisive motivation or influence shaping management strategy (see also Batstone *et al.* 1987: 210–19). In Chapter 3 we argued that labour process theory ran the risk of misinterpreting the nature of managerial intentions when they seek to use new technology to increase control. While the pursuit of 'operational control objectives' may involve attempts to reduce the need for skilled or informed human intervention, managerial intentions cannot therefore be reduced exclusively to a desire to deskill labour. In fact, other concerns may be more significant and in certain circumstances require a 're-skilling' or 'up-skilling' of work. In the cases reviewed from our own research it was also revealed that lower-level managers may have little in the way of corporate or strategic guidelines in relation to issues such as job content, work organisation and supervision. As a result their decisions – or non-decisions – could frustrate the achievement of the overall objectives behind the introduction of new technology.

It is also important to distinguish between the objectives of managers at different levels in the hierarchy if the nature of management strategies is to be fully understood. In an interesting and persuasive argument Buchanan (1986) points out that the structural position of managers within organisations can support particular sets of vested interests or 'stakes'. 'Stakes' vary between management functions and levels and may not be consistent with overall organisational objectives. In particular, lower-level managers may be more concerned with personal objectives defined in terms of career aspirations within the organisation's control and reward systems. Thus individuals may be more inclined to make decisions which are best for their careers, but not always best for the organisation. 'It may be,' suggests Buchanan, 'that for many middle managers, personal aspirations have a more powerful influence over decisions concerning technical change than organisational objectives. (1986: 79).

Buchanan argues that the persistence of narrow and personal outlooks and the tendency for British managers to fail to adopt a long-term, strategic and organisational perspective when introducing new technology, is indicative of the short-term tactical outlook engendered within British industry and the British system of management education which fails to encourage alternative thinking (1986: 80). To this failure might be added a neglect of the human and behavioural elements which comprise much of organisational life. It is significant that two recent and influential reports on management education and training in Britain have criticised the pragmatism and lack of professionalism of British managers (see Handy 1987; Constable and McCormick 1987). If, as these reports advocate, more managers at all stages in their careers are to be exposed to education in Business Schools and the like, then this suggests that important opportunities to challenge and reconstruct what Peter Anthony (1986) has called the existing moral foundations of management authority may present themselves. However, as Anthony argues, this will require management educators themselves to place a far more critical emphasis on behavioural and social aspects of managerial roles and rather less on the neutral application of quantitative techniques in financial analysis, marketing and decision analysis conventionally 'deemed to be the equipment necessary to a good manager' (1986: 120).

Measuring the effective use of new technology

A further factor undermining managers' ability to use new technology effectively may be that the traditional productivity and performance indicators that are used in justifying and evaluating current practices may not be appropriate when new computing and information technologies are introduced. For example, Boddy and Buchanan argue that the conventional approach to productivity – usually understood in terms of ratio of inputs to outputs related to a particular activity – is a useful but limited guide to the overall effectiveness of the use being made of new technology (1984: 234). Thus, whereas significant productivity gains can be demonstrated when new technologies are introduced – such as increases in the number of words typed per minute by a word-processor operator or the reductions in the time taken by drafting staff to produce a standard drawing on a CAD system compared to a manual drawing board – these do not necessarily mean that the technology is being used most effectively. Rather, these measures of efficiency refer to only limited areas of use of the technology, in these cases within a centralised word-processing centre or drawing office. As Boddy and Buchanan point out, these 'measurable aspects of performance ... may be emphasised at the expense of other equally important but less measurable factors, such as the quality of the end product'; and ' "local" measures of productivity, in one part of the system, take no account of productivity in related operations' (1984: 234; see also McLoughlin 1987).

Similarly, in a review of recent research on the adoption of microelectronics, Peter Senker concluded that the justification and evaluation of new systems tended to reflect traditional investment appraisal practices in Britain. Typically proposals for the adoption of new equipment or systems are initiated by the functional area or department in which it is to be utilised. In line with conventional investment appraisal techniques, justifications and evaluations are made on the basis of the reductions in direct labour costs in the departmental function concerned. Cost benefits are then estimated against the cost of the proposed capital investment. However Senker argues that there are a number of other criteria which are generally not given so much weight, but which may be critical if the effectiveness of new computing and information technologies is to be adequately assessed. These include cost-saving considerations apart from direct labour-saving such as overheads, materials, indirect labour and work-in-progress; benefits and costs in functional areas outside of that in which the equipment or system is installed; and the need for new skill requirements and forms of work organisation in order to operate the technology effectively (1985a: 157–8). In practice, the tendency for managers to justify their investments in terms of the benefits of reduced labour costs means that lost benefits in terms of the other criteria, in particular the under- and ineffective utilisation of human resources, may be ignored.

More generally it has been suggested that measures of productivity and efficiency are not objective and scientific, but ambiguous and imprecise in character and open to selective use and interpretation (Wilkinson 1983: 82–3; Buchanan and Boddy 1983: 250–1; Senker 1985b). For example, where computing and information technologies are concerned, they may be introduced into areas of an organisation's operations, such as the drawing office or administrative work, which have hitherto not experienced significant capital investment. In these circumstances a high level of managerial estimation, if not guesswork, may be involved to 'get proposals past' senior and financial management. This may mean that criteria used to justify the purchase and mode of operation and evaluate the performance of new equipment may be inappropriate measures of effectiveness, which may be being deployed for what are at least partly political reasons.

Wilkinson goes as far as to suggest that efficiency claims made by managers in such circumstances may be ideological justifications for what are entirely political decisions (1983: 84). Perhaps a less extreme view is that in many instances efficiency is a far more 'fuzzy' concept than is usually claimed and one which, if conventional approaches are allowed to prevail, may provide an unwarranted justification for traditional management practices. How, though, may current management assumptions and thinking be challenged?

Changing management practices: the role of personnel specialists

In Chapter 3 it was noted that personnel and industrial relations specialists might have been expected to play a leading role as 'change agents' in developing innovative responses to the human and organisational issues associated with the introduction of new technology. This is what Thomason (1981) has referred to as the 'organisational consultant' role of the personnel manager. However, the evidence reviewed in this book has revealed that normally personnel specialists have had little influence over the process of technological change, whether in relation to strategic decisions or over such issues as work organisation and supervision. A further reason for the continuing predominance of traditional assumptions and values, therefore, may well be ᴛthe absence within the management function of an alternative voice pointing to the human and organisational implications of technological change. If conventional management thinking is to be challenged from within organisations this suggests that a particular responsibility for promoting alternatives could rest with personnel and industrial relations specialists.

The challenges and problems confronting personnel managers have been spelt out as follows by Buchanan:

> The first challenge for personnel managers in Britain over the next decade will be to convince colleagues in other functions that human resources are a key asset in competitive strategy. The second challenge will be to introduce forms of work design and related employment policies that develop and sustain employee performance and commitment. The constraints on personnel management will be the continued scepticism and the financial and technical preoccupations of line management, and the rapid rush of process and product innovation which apparently leave little time to consider the 'soft' human and organisational issues.
>
> (forthcoming)

Similar sentiments underlie Wilkinson's view that personnel and training managers have to overcome the syndrome of 'engineers supply the technology, we deal with the consequences' (1983: 97).

In Chapter 3 it was suggested that the personnel involvement in technological change, where it does occur at present, tends to be reactive and limited to dealing with problems that arise during or after the implementation of new technology when many of the options or alternatives have already been closed off. Clegg and Kemp (1986) advance an alternative model to this 'sequential approach' to management decision-making. It provides for a more proactive role for personnel specialists in the strategic decisions regarding change and enables human design aspects to be dealt with in parallel with technical design. This requires the overall philosophy and objectives for both technical and human aspects of change to be established by a senior management team. Decisions on the organisation structures and work roles needed to control, manage and operate the new technology are then delegated to a human design group working parallel to a technical design group (see Figure 8.1). Finally, detailed planning tasks are delegated to relevant functional specialists, for example to devise training programmes for the new work roles that are to be created. In this way, they argue, personnel specialists might take up a more

Figure 8.1   Methods of introducing new technology

(a) Sequential method of introducing information technology

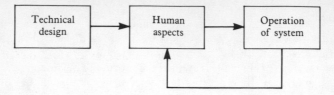

(b) Parallel method of introducing information technology

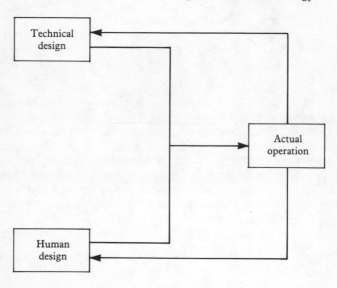

Source: Clegg and Kemp (1986)

strategic and central role in both the management of change and the actual use made of new technology once operational.

Trade unions and workforce participation

On past evidence the kinds of changes in management attitudes and assumptions that appear to be necessary if new technology is to be used effectively are unlikely to be brought about totally from within. In this context it is important to recall the failure of trade unions to concern themselves sufficiently with questions surrounding the strategic and control issues highlighted by new technology. In Chapter 4 the popular but inaccurate view that trade unions are a principal barrier to technological change was challenged. However, it was also evident that union attempts to influence such change via new technology agreements and national economic planning have been largely ineffective. Even where some success has been achieved, this has been over traditional employment issues, rather than strategic or control issues.

It can also be suggested that union policies have tended to be 'gender blind' and oriented towards the protection of the interests of male job-holders, as seen for example in the case of Smiths Crisps in Chapter 4. The problem for trade unions, therefore, is to find more effective means of broadening the range of issues over which they negotiate, and of finding ways in which sectional divisions, between skilled and unskilled and male and female workers, might be overcome. Indeed, by placing more emphasis on resolving these issues it could be that trade unions might also play a role in encouraging changes in managerial attitudes and practices in the future.

Levie and Williams (1983) have outlined three possible trade union strategies for influencing technological change. The first is essentially that which has been pursued by trade unions in Britain in the post-war period and involves confining bargaining to major employment issues such as wages and job security. This implies conceding most of the wider areas of decision-making over the introduction of technology to management. The second option is to seek to bargain over the full range of effects of technological change in the workplace, including both traditional employment issues and the more directly work-related control issues. Levie and Williams describe this as 'external influence', since it does not attempt to become involved in management decisions over change but seeks to bargain over them either before, during or after the process of strategic decision-making. Like the first option this is initially a reactive approach, but unlike the first it also places a high premium on the development of credible alternatives to management proposals. The later bargaining is left in the process of change, the more difficult it becomes to achieve union objectives because management will already have taken major strategic decisions such as which type of equipment or system to install. The third option is direct involvement in strategic issues through joint working parties and the like and with trade union representation in the system design process. This is described as 'internal influence'. The advantage here is that as a party to the decision-making process unions could play a proactive role and directly influence the 'impact' of change on the workforce.

However, approaches such as this involving workforce participation in strategic decision-making are not without their problems. In Britain, internal influence of this type, and indeed the general extension of industrial democracy by these means, has been regarded in trade union circles with suspicion. There is still a widespread belief that union representatives on joint committees or even company boards would have a negligible influence, but that their presence would allow management to claim union support for its decisions (see Batstone *et al.* 1983). According to Levie and Williams the problem is to 'maintain an independent union perspective through such a participatory process' (1983: 279). In the case of technological change this would require: the development of independent views on new equipment and systems to enable representatives to present alternatives to management proposals where appropriate; procedures to mobilise union support and maintain the accountability of representatives on participatory bodies; organisational and other resources to facilitate these developments (1983: 279). Even where these conditions were met, Levie and Williams argue that, internal influence would have to be linked to external influence to make sure that decisions reached through participatory frameworks were then subject to collective bargaining. They conclude, 'this would have to involve changes in the method of introducing microprocessor-based production and information systems – with each stage of decision-making and design open to amendment' (1983: 280).

We would argue that the strategy developed by the local ACTT branch in the ENG case study reported in Chapter 3 would be an excellent role model for trade unions to follow in this respect (see Clark *et al.* 1984). Also relevant here are recommendations made by Stephen Jary on the basis of his more recent research on improving the trade union responses to new technology at workplace level (see Chapter 4 for a summary of some of this research). His findings support the conclusions reached by Batstone and Gourlay (1986) (see also Batstone *et al.* 1987) that the level and range of bargaining over new technology is related to the degree of sophistication of the workplace trade union organisation and the degree of external integration between it and the wider union structure. Jary's argument is that improvements in the professionalism of trade union response could be achieved by increasing external integration, which in turn could lead to more sophisticated responses on the part of management to the issues highlighted by technological change.

Central to increasing external integration, suggests Jary, is the role of the full-time officer in servicing shop stewards and in educating stewards and members on new technology and bargaining techniques and options. However, this also implies a prior union investment in the training of its full-time officers, and perhaps the development of specialist skills and expertise in particular areas such as technological change. Wilkinson suggests that the single most useful measure that unions could take in this respect would be to 'make frequent use of thorough research into the specific implications of all new technology introduced into plants'. This might allow negotiators to 'justify alternative production machinery in order to influence the direction of change before certain constraints are imposed' (1983: 100).

None of these proposals, however, have much to say about gender issues and ways of both increasing the participation of women and ensuring that they gain an equal share of the benefits of technological change while not being subjected to a disproportionate share of its negative effects. One of the major problems here is that despite their growing participation in the labour force women have a low involvement in trade union affairs, even in unions where women make up the majority of the membership. The task of involving women in bargaining over technological change, therefore, involves the additional consideration of making the union itself a more acceptable vehicle for the representation of women's interests. Murray (1987) points to the example set by some Scandinavian unions as one that might be followed by their British counterparts. For example, unions concerned with the introduction of new technology there have developed informal 'study circles' held at the workplace which provide a forum conducive to the discussion of collective issues by women. This contrasts with the formal branch meetings of many British unions, typically held out of work hours in public houses and organised by male-dominated committees.

However, it has been argued that policies designed to encourage equal opportunities may not be sufficient to avoid women workers becoming the focus for many of the negative effects of technological change in the areas of employment in which female labour is concentrated, for instance in secretarial and clerical work, or to ensure that any new opportunities and jobs created in new technical occupations are equally open to them. According to Cynthia Cockburn (1986) 'opportunity is not enough', since the identities which are attached to performing work roles with a technical content are perceived both by men and women as peculiarly male. On the other hand, women's work and domestic roles are traditionally associated with low levels of technical literacy and competence. Women offered the 'opportunity' to re-train for work using new technology are discouraged from seeking to acquire technical competence by the knowledge that to do so will mean they will be considered 'unfeminine and unloveable'. According to Cockburn, equal opportunity seen in this way is an inadequate formula for women since 'it will mean little more than the "masculinising" of a handful of women unless more profound changes occur at the same time' (1986: 187).

On this analysis more fundamental changes would be needed, including a redefinition of the relationship between paid work and unpaid domestic work and a far greater acceptance by men of domestic responsibilities. Although unions such as NUPE and the GMB have recently made significant moves towards developing policies designed to appeal more directly to women workers, these are still a long way from the kinds of changes implied in Cockburn's analysis.

One major argument against all these suggestions is that the legal, political and economic environment and the weak labour market position of trade unions in the 1980s undermine any attempt by unions to influence management decision-making. In this view any internal policy and organisational changes are likely to be ineffective if not merely cosmetic as long as the sources of trade union weakness prevail. However, it might reasonably be countered that even if conditions more conducive to the extension of collective bargaining and increases in trade union influence had prevailed during the 1980s, the 'blind spots', organisational deficiencies, and deep-seated structural and ideological changes that have been taking place in the economy and society would still have been evident (see Moore and Levie 1985; Clark 1987). In this sense the idea that unions should 'sit out the recession' and expect 'normal service to be resumed' once a Labour Government is elected is at best shortsighted.

Nevertheless, the importance of a supportive political and economic environment has been demonstrated in one of the few new developments during the 1980s which has offered trade unions the possibility of extending their influence over technological change. This has been provided by the various 'innovation initiatives' developed by Labour-controlled local authorities – in particular the Greater London Enterprise Board (GLEB) established by the now defunct Greater London Council (GLC) and similar schemes in Sheffield, Merseyside and the West Midlands.

The GLEB innovation initiative is particularly significant since its antecedents lie in the attempts by shop stewards at Lucas Aerospace during the 1970s to develop alternative corporate plans to those proposed by the Company. At the time these were seen as a new departure in the development of industrial democracy (see Wainwright and Elliott 1982; Cooley 1985). The GLEB was established in 1984 as part of the GLC's commitment to 'reverse industrial decline by establishing a capacity to intervene in key aspects of the local economy' (Labour Party manifesto for Greater London Council elections, quoted from Open University 1986b: 87). The primary aims of GLEB were to: '(1) create and improve the quality of jobs in London; (2) regenerate London's economic base by investing in London's industry; (3) widen the control that Londoners have over their economic future, particularly at work' (GLEB Corporate Plan 1984–5, quoted from ibid: 88).

Subsequently GLEB has provided investment funds for new products and companies and established 'technology networks' in collaboration with London polytechnics, including one concerned with the development of human-centred production systems (see below). Significantly, the appraisal of investment proposals included the criterion of social costs and benefits and looked in particular for a willingness on the part of owner/managers to enter into planning agreements with trade unions to cover such things as the introduction of new technology and work organisation. However, the demise of the GLC in 1986 meant that GLEB's activities since that time have been severely curtailed (for an assessment see Elliott 1986).

Perhaps the most fundamental problem facing trade unions in dealing with new computing and information technologies, however, is the pattern of innovation itself. In Chapters 1 and 3 reference was made to a distinction drawn by Paul Willman between those firms who have adopted microelectronics in their new technology products, and those who have adopted the products – such as CNC machine tools, CAD systems, word processors, computer and telecommunications systems and so on – in their production processes. Firms adopting new technology in their products (product innovation) do so, Willman argues, as part of a business strategy designed to maximise performance and improve market share. Firms adopting new technology in their processes (process innovation) tend to do so in order to cut costs, with implications in particular for job security.

In Britain many of the leading suppliers of new technology products are foreign owned multi-nationals. In such companies employees might impose conditions on new product innovation without fear of job loss. In contrast many of the firms now adopting new technology in their processes are in mature industries with high union densities where the alternative to accepting management plans is often closure and redundancy. The paradox for trade unions, as Willman points out, is that in the areas where their influence over management may be weakest they are best organised, whilst in the areas where the potential to challenge management proposals is greatest they are, if recognised at all, normally weak and unable to intervene (1986: 256–9).

Towards the computer-aided workforce: some alternatives

Notwithstanding the constraints and difficulties facing managers and unions in re-thinking their current approaches, what are the alternatives in work design and supervision that might be considered in promoting the use of new technology where it can complement rather than replace the need for skilled and informed human intervention? At least two possibilities have emerged in recent research. The first is concerned with the relationship between technology and the human operator, and in particular the idea that system design might be 'human-centred'. The second has relates to the design and control of work itself, and in particular the implications for workers and their supervisors which might follow from a recognition of the need – noted in some of the cases discussed in Chapter 7 – to delegate more control to the workgroup.

Human-centred system design

One of the foremost advocates of the idea of a human-centred approach to system design in Britain has been Howard Rosenbrock, Professor of Control Engineering at the University of Manchester Institute of Science and Technology (UMIST)[1]. With funding from the Joint Committee of the SERC/ESRC, Rosenbrock's team of researchers, consisting both of social scientists and engineers, has attempted to develop a human-centred design for a manual data input (MDI) computer numerically controlled (CNC) machine tool (see Chapter 6 for a description of MDI–CNC).

One of Rosenbrock's co-workers contrasts the new approach adopted with conventional ways of viewing system design as follows:

> The conventional, technocratic approach to system design involves designers in the practice of man–machine comparability – functional requirements being realised with respect to the technological state of the art, where man (sic) takes over those functions that are technically not yet solved . . . The alternative, or human-centred approach to man–machine system design rejects the notion of man–machine comparability and focuses instead on how they may complement each other. In this view, men and machines help each other to achieve an effect of which each is separately incapable.
>
> (Corbett 1985: 202)

The key point is that the effectiveness of this approach does not rely on reductions in the cost of labour input to the machining process. Rather,

> The efficiency of a human-centred system is based on the complementarity of operator and machine. Because of unforeseen disturbances that may enter the system, the operator must be able to control all tasks that contain choice-uncertainty via an interactive interface. However, an operator cannot control a system without comprehending its functioning. A system should support the operator's model of its functioning (schema) so that knowledge that is needed during infrequent task activity is obtained during general activity.
>
> (1985: 211)

Corbett illustrates the difference of these two approaches in practice in the software design for a human-centred MDI–CNC lathe. In principle three types of work task are associated with machining:

Type A tasks: Those tasks which can be performed only by the machine operator.
Type B tasks: Those tasks which can be performed by a machine operator but can also be performed automatically and/or away from the machine.
Type C tasks: Those tasks that cannot be performed by the machine operator.

The flexibility of the software which is used to program microprocessor controls on CNC machines can be used in the design of most Type B tasks[2]. The conventional approach seeks to automate every Type B task so that the computer program contains all the information necessary to describe the components of the machining process and their interaction. Where errors exist in the program these are corrected by the programmer away from the machine. This type of system design allows for no interface between the operator and the machine, 'leaving the operator subordinate to the machine' (1985: 202) – hence the locked cabinets noted in studies of CNC machine use (see Chapter 6). The 'human-centred' approach, on the other hand, seeks to find ways in which the skill of the operator in making decisions over how Type B work tasks are to be executed can be complemented by computer controls, thus enabling the operator to resolve 'unforeseen disturbances' (for example variations in tool wear, poor casting, poor surface finish, which cannot be detected by the machine) in the system at the point at which information about them is generated – that is, at the machine itself.

In order to allow operators to use their skills and discretion to intervene in this way and correct mismatches between the software programming and the actual components and interactions of the machining process, an interface has to be designed which in effect allows the operator and computer to 'talk to each other' on equal terms. The CNC lathe used in the UMIST experiments calculates cuts within a variety of physical constraints enabling the machine software to calculate four variables: the speeds, feed-rates, depths and number of cuts (1985: 203). Adopting a human-centred approach to designing the interface between the machine software and the human operator involves providing operators with a display at the machine of computer-calculated data on these four variables. Operators are able on the basis of their own experience and skills to change any of the calculated values they regard as incorrect. The remaining values are then re-computed in the light of any manual changes to check that the redefined cutting conditions are within the physical constraints under which the machine can operate. As a result, the control functions involved in producing a finished workpiece to required tolerances and quality can be 'performed concurrently by operator and computer' (1985: 209).

For Marxist critics, whilst it challenges managerial assumptions, such an approach does not necessarily involve a challenge to the economic interests of the employer. Such critics appear to argue that human-centred systems can be brought about only in non-capitalist modes of production. However, advocates of the human-centred approach such as Rosenbrock, argue that it challenges dominant managerial and engineering practices in both capitalist *and* state socialist societies (see Rosenbrock 1985a; 1985b). Smith points out, in this guise the human-centred approach is 'primarily anti-*managerial* rather than anti-*capitalistic*' (1987: 161, original emphasis). Whether this is seen as a weakness or an advantage of the human-centred approach is clearly dependent to some extent upon the political position of the observer. In fact, it might be more accurate to characterise human-centred systems design as 'anti-technocratic' since it is presented as an alternative to the narrow technology-driven perspective endemic in the 'object-oriented' occupational culture of the technical specialist (see Rosenbrock 1985a; McLoughlin 1983a).

One apparent criticism of the UMIST approach has been the absence of continuous worker and trade union involvement in the design process (Rosenbrock

1985b). However, according to Rosenbrock, this reflects the practical problems faced by human-centred system designers. Previous attempts at developing human-centred system design methodologies had found that the full participation of end users in decisions during system design is hampered by the fact that the development of user competence usually lags behind the actual design of the system. This means that there is only a short 'window' during the design process in which end users are sufficiently knowledgeable about the system to make an effective contribution and in which sufficient decisions about the technical design of the system remain for their views to have influence.

The UMIST project involved periodic in-depth interviews with machine operators and feedback from them on software simulations. There was a danger that the design would be influenced only by the system designers if the input of human operators could not be utilised. According to Rosenbrock, a more effective way of involving workers to the point where they are able to give definitive conclusions on system design would require them to use the technology under 'industrial conditions' for at least six months. The problem with this proposal is that by this time the scope for choice in the design would be drastically reduced (1985b: 5).

Even if such difficulties can be resolved, problems of redesigning work around the technology to allow operators to exercise control over the machining process still remain. In practice this will mean challenging the existing formal divisions that exist in many machine shops between machinists and programmers (see Chapter 6). According to Bryn Jones the effective use of human-centred systems of the type proposed by the UMIST team:

> ...would require more fundamental changes in occupational organisation in order to transcend these differences and to create an atmosphere in which devolved decision-making was seen as desirable by management and different occupational groups.
>
> (1982b: 9)

In Jones's view, the kinds of checks and safeguards that could be built into human-centred software to prevent human errors leading to costly damage to workpieces might convince production managers of the technical and commercial merits of devolving control to the operator. However, the major obstacles would lie in the policies of craft and ex-craft trade unions which seek to prevent the dilution of occupational boundaries by defending the skilled status of job holders while the skill content of the work itself is transformed. As Jones notes, and as was seen in Chapter 6, such policies may have had some relevance to the immediate past of the present phase of industrial automation, but current developments suggest that the next stage will render the defence of such skill labels highly questionable.

Alternative forms of work design and supervision

As the above comments indicate, the second area of possible alternatives lies in the design and supervision of work itself. The principal question here is whether the various work redesign techniques that have been developed in the past might enjoy a new and perhaps more successful lease of life if utilised when computing and information technologies are introduced. As Buchanan puts it:

> Sophisticated, flexible, expensive equipment needs sophisticated, flexible,

expensive people to operate it. The effective and safe operation of these new technologies requires very careful attention to work design.

(1985a: 27)

What then are the alternative choices that may be available?

Many of the alternatives owe their origins to the various work redesign initiatives that were developed in North America and continental Europe during the 1960s and 1970s but which never took hold in Britain to any great extent (see Buchanan forthcoming, for a brief discussion of the literature; also Bailey 1983; Child 1984; Littler and Salaman 1984). These alternative approaches fall into two categories, but the objective of both is to upskill various aspects of job content. The first focuses on ways in which individual jobs can be altered to either extend the range of tasks carried out (for example job rotation or job enlargement) or expand the scope for job-holders to exercise discretion and expertise (for example job enlargement). The second focuses more broadly on the relationships between jobs, rather than simply on their individual content, and explores the possibilities for upskilling that can be pursued by redesigning groups of jobs. The most significant and potentially powerful of these techniques is the idea of 'autonomous work groups'. This involves work being designed so as to allow the task complexity and discretionary content of individual jobs to be increased and for work groups to be allocated the responsibility for deciding how work is to be organised and executed in particular areas of production.

The advantages of alternative work design were often expressed in the past in humanitarian and philanthropic terms as an antidote to the alienating and dehumanised work conditions and experience that have often accompanied technological change. However, it is important to note that in reality the attractiveness of alternative work design to management has been driven by concerns related more directly to the profitability and efficiency of the enterprise. Certainly, many of the managerial experiments in work design in the 1970s appear to have been motivated by a concern to overcome the effects on production of absenteeism, unofficial strikes and low motivation rather than a commitment to make work more satisfying *per se* (Bailey 1983; Child 1984; Buchanan forthcoming). From a management viewpoint, therefore, any new options opened up by computing and information technologies are likely to be of interest first and foremost because they improve the effective use of the technology itself and thereby the performance of the organisation. As Boddy and Buchanan note, any 'concern with work organisation in the late 1980s and into the 1990s' will be motivated in part by 'pressures arising from stiffer trading conditions in domestic and international markets' and the realisation that 'if the competition can buy the same technology, then competitive advantage must depend on organisational innovation' (1986: 105, 228).

The informal emergence of 'team autonomy' noted in some of the case-study research in Chapter 7 points to the potential for the more formal development of approaches to work design based around autonomous work groups. Buchanan (1985b: 464) argues that the optimum form of job content and work organisation is one that gives workers:

1 more and faster information feedback on performance;
2 meaningful, interesting, and challenging goals;

3 control over the workflow;
4 discretion over methods of task allocation;
5 opportunities to develop skill and knowledge through work.

This is similar to the idea of 'intrapreneurial groups' advanced by Norman Macrae (1982). Intrapreneurial groups are autonomous groups within organisations consisting of around ten members and allocated a budget to provide services to the wider organisation. Buchanan claims word-processing centres might be most effectively organised in this way since they could act as 'minifirms' supplying services to their 'customers' elsewhere in the organisation. In the case of computer-aided machine tools, Jones argues that alternative forms of work organisation might involve a single flexible category of 'machine-programmers' where experience of machining work would be combined with short training courses in basic information processing and work planning techniques (1982b: 9).

In the past, work redesign schemes have tended to fail because they have not recognised implications for other parts of the production process such as management control, supervision and payment systems (Child 1984). Buchanan argues that, to be successful, work redesign schemes have to be 'applied as part of an integrated employment package (Buchanan forthcoming). The importance of combining changes in job content and work organisation with developments in payment systems has been emphasised by numerous writers (see for example, Wilkinson 1983; Child 1984; Buchanan forthcoming; Martin 1988). For example, the creation of integrated grades of machinist–programmers would also require the creation of an integrated payment system which recognised both changes in the pattern of required skills and the need for smooth progress from jobs which predominantly involve machining to those which place greater emphasis on technical programming tasks.

Computing and information technologies also break the direct link between individual effort and output, making traditional piecework payment systems obsolete because the quantity and quality of output is based on the performance of the work group as a whole. In such circumstances, argues Buchanan, 'payment systems may have to be designed around the need to encourage flexibility and adaptability'; and 'the more visible the impact of individual or group on the organisation as a whole, the more appropriate financial participation becomes as the basis of incentive schemes' (Buchanan forthcoming). This suggests that employee shareholding, profit sharing and above all team bonuses may be among the more appropriate forms of reward to accompany work redesign on the basis of autonomous work groups.

The need to relate changes in job content and work organisation to management control and supervision has been noted by a number of writers (see for example Child 1984; Dawson 1986; Burnes 1987). Child argues that such changes tend to have a 'shunt' effect upon the distribution of authority and the nature of hierarchical control. and that the devolution of control has implications for the roles of first-line supervisors and line managers (1984: 42). Aspects of this problem have been evident in the research in which we have been involved with Patrick Dawson. This work indicated that the choices opened up by new technology may be better understood if work group leaders, first-line supervisors, and lower level line-managers are seen as part of a 'supervisory system' (Thurley 1966; Thurley and

Wirdenius 1973) consisting of a hierarchy of roles. At one end are mixed operator/supervisory roles, sometimes referred to as 'working supervisors', whilst at the other are jobs designated as managerial but which incorporate supervisory tasks to form 'second-line' supervisory roles. In the middle are formally designated or 'pure' first-line supervisory roles (see Dawson 1986; Dawson and McLoughlin 1986).

This means that the range of choices open to managers may require a complete review of supervisory responsibilities when new technology is introduced (see Dawson 1986; Dawson 1987a; Dawson and McLoughlin 1987; Clark *et al.* 1988). For example, devolving decisions to the shop floor will create new responsibilities for 'working supervisors' as team leaders to co-ordinate work group activity. Similarly, whilst 'mechanical controls' erode the need for some aspects of direct personal control to be exercised by first-line supervisors, new 'composite roles' may be needed combining new tasks with existing responsibilities. Finally, the new information processing and control capabilities of the technology may enable the creation of new second-line supervisory and lower line-management roles based around 'information management' tasks. In short using new technology effectively could entail a complete redefinition of supervisory management. As David Buchanan rightly observes:

> The advantages of new technology may be lost if it is applied with traditional management values and assumptions. The effective use of new information and computing technologies may be dependent on new forms of work organisation. This may in turn necessitate a reconstruction of the role of management, particularly at lower levels of line management.
>
> (1983: 79)

Concluding comment

A great deal has now been written about the potential and actual effects of new computing and information technologies at work. In this book an attempt has been made to draw together and evaluate hard research evidence on the process of technological change as it has occurred in the workplace. The growing body of such evidence suggests that management and unions still have a lot to learn in devising the most effective methods of introducing and working with new technology. It is our hope that this book will contribute to this learning process both for students and practitioners.

Notes

1 Research into 'human-centred' system design has a much higher profile and is more established in Scandinavia, in particular in Sweden where the Swedish Centre of Working Life has been working on various projects to democratise the design of computer-based systems since the late 1970s. For a discussion see Gill 1985: 141–60.
2 Type B tasks include loading and unloading the machine; machine set-up; making jigs; ensuring tools and workpieces are available; determining cutting programmes; conducting test runs; storing programmes; production control; work analysis; and planning. Type A tasks are associated with running the batch, and Type C tasks the design of the artefact itself. See Corbett 1985: 204.

APPENDIX

Figure A.1 Main difficulties of microelectronics users

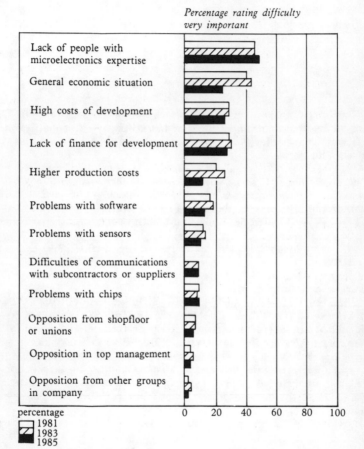

All UK factories

*Percentage rating difficulty
very important*

Source: Northcott (1986)

Bibliography

ACARD (Advisory Council for Applied Research and Development) (1979). *Technological Change: Threats and Opportunities for the United Kingdom*. London, HMSO.

Albury, D. and Schwartz, J. (1982). *Partial Progress: The Politics of Science and Technology*. London, Pluto Press.

Anthony, P. D. (1986). *The Foundation of Management*. London, Tavistock.

Arnold, E. and Senker, P. (1982). *Designing the Future: The implications of CAD interactive graphics for employment and skills in the British engineering industry*. Occasional Paper No. 9. Watford, Engineering Industry Training Board.

Arnold, E. *et al.* (1982). Microelectronics and women's employment. *Employment Gazette*, September, 376-7.

Atkinson, J. (1984). Manpower strategies for flexible organisations. *Personnel Management*, August, 28-9.

Atkinson, J. and Meager, N. (1986). Flexibility – just a flash in the pan? *Personnel Management*, September, 26-7.

AUEW–TASS (1979). *New Technology: a guide for negotiators*. AUEW–TASS.

(1985). *Shiftwork and the Whitecollar Worker*. AUEW–TASS.

Ayers, R. and Miller, S. (1985). Industrial robots on the line. In Forester, T. (ed.), *The Information Technology Revolution*. Oxford, Basil Blackwell, 273-83.

Bailey, J. (1983). *Job Design and Work Organisation*. London, Prentice-Hall.

Bain, G. (ed.) (1983). *Industrial Relations in Britain*. Oxford, Basil Blackwell.

Baldry, C. and Connolly, A. (1986). Drawing the line: computer aided design and the organisation of the drawing office. *New Technology, Work and Employment*, **1** (1), 59-66.

Bamber, G. (1980). Microchips and industrial relations. *Industrial Relations Journal*, **11** (5), 7-19.

Banking World (1986). Cashless shopping – vertical take off. January, 28.

(1986). Forces to be reckoned with. March, 36-7.

(1986). Study in super-smart cards. December, 31.

(1986). Annual ATM Survey. November, 50-60.

(1987). The renaissance of branch banking. April, 28-9.

(1987). Annual ATM Survey. November, 49-57.

Barker, J. and Downing, H. (1985). Word processing and the transformation of patriarchal relations of control in the office. In Mackenzie, D. and Wajcman, J. (eds), *The Social Control of Technology*. Milton Keynes, Open University Press, 147-64.

Barnett, C. (1986). *Audit of War*. London, Macmillan.

Barron, I. and Curnow, R. (1979). *The Future with Microelectronics*. London, Frances Pinter.

Batstone, E., Ferner, A. and Terry, M. (1983). *Unions on the Board: An Experiment in Industrial Democracy*. Oxford, Basil Blackwell.

(1984). *Consent and Efficiency: Labour Relations and Management Strategy in a State Enterprise*. Oxford, Basil Blackwell.

Batstone, E. and Gourlay, S. (1986). *Unions, Unemployment and Innovation*. Oxford, Basil Blackwell.

Batstone, E. *et al.* (1987). *New Technology and The Process of Labour Regulation*. Oxford, Clarendon Press.

Beardwell, I. J. (1987). *New Technology and Bargaining Strategy: the Case of the Civil Service Unions*. Occasional Paper, School of Industrial Relations and Personnel Management, Kingston Business School, Kingston Polytechnic.

Bedeian, A. (1980). *Organisations: Theory and Analysis*. Illinois, The Dryden Press.

Beechey, V. (1982). The sexual division of labour and the labour process. In Wood, S. (ed.), *The Degradation of Work?: Skill, Deskilling and the Labour Process*. London, Hutchinson, 54–73.

Bell, D. (1980). The social framework of the information society. In Forester, T. (ed.), *The Microelectronics Revolution*. Oxford, Basil Blackwell.

Benson, I. and Lloyd, J. (1983). *New Technology and Industrial Change*. London, Kogan Page.

Bessant, J. (1982). *Microprocessors in Manufacturing Processes*. London, PSI.

(1983). Management and manufacturing innovation: the case of information technology. In Winch, G. (ed.), *Information Technology in Manufacturing Processes*, London, Rossendale.

Bessant, J. and Haywood, B. (1985). *The Introduction of Flexible Manufacturing Systems as an Example of Computer Integrated Manufacturing*. Innovation Research Group, Brighton Polytechnic, October.

Blauner, R. (1964). *Alienation and Freedom*. Chicago, University of Chicago Press.

Blumberg, M. and Gerwin, D. (1985). Coping with advanced manufacturing technology. In Rhodes, E. and Wield, D. (eds), *Implementing New Technologies: Choice, Decision, and Change in Manufacturing*. Oxford, Basil Blackwell, 352–60.

Boddy, D. and Buchanan, D. A. (1984). Information technology and productivity: myths and realities. *Omega* **12** (4), 233–40.

(1986). *Managing New Technology*. Oxford, Basil Blackwell.

Braun, E. (1986). Government policies for technology. In Roy, R. and Wield, D. (eds.), *Product Design and Technological Innovation*. Milton Keynes, Open University Press, 210–18.

Braverman, H. (1974). *Labor and Monopoly Capital: The Degradation of Work in the Twentieth Century*. New York, Monthly Review Press.

Bright, J. (1958). *Automation and Management*. Boston, Harvard University Press.

Buchanan, D. (1979). *The Development of Job Design Theories and Techniques*. Farnborough, Saxon House.

(1983). Technological imperatives and strategic choice. In Winch, G. (ed.), *Information Technology in Manufacturing Processes*. London, Rossendale, 72–80.

(1985a). Canned cycles and dancing tools: who's *really* in control of computer aided machining? Paper presented at ASTON/UMIST Conference on the 'Organisation and Control of the Labour Process'. Manchester, April.

(1985b). Using the new technology. In Forester, T. (ed.), *The Information Technology Revolution*. Oxford, Basil Blackwell, 454–65.

(1986). Management objectives in technical change. In Knights, D. and Willmott, H. (eds), *Managing the Labour Process*. Aldershot, Gower, 67–84.

Forthcoming. Principles and practice in work design: current trends, future prospects. In Sisson, K. (ed.) *Personnel Management in Britain*. Oxford, Basil Blackwell.

Buchanan, D. and Boddy, D. (1983). *Organisations in the Computer Age: Technological Imperatives and Strategic Choice*. Aldershot, Gower.

Buchanan, D. and Huczynski, A. (1985). *Organisational Behaviour*. London, Prentice-Hall.

Buchanan, D. and McCalman, J. (1988). Confidence, visibility and performance. In Boddy, D. McCalman, J. and Buchanan, D. (eds.), *The New Management Challenge*. London, Croom Helm, 17–29.

Bullock (1977). Committee of Inquiry on Industrial Democracy. *Report*. Cmnd 6706. London, HMSO.

Burawoy, M. (1979). *Manufacturing Consent*. Chicago, University of Chicago Press.

Burnes, B. (1987). New technology and the role of supervisors. *Employee Relations* **9** (4), 9–13.

Business Week. (1985), Software: the new driving force. In Forester, T. (ed.), *The Information Technology Revolution*. Oxford, Basil Blackwell, 27–44.

Bylinsky, G. and Hills Moore, A. (1985). Flexible manufacturing systems. In Forester, T. (ed.), *The Information Technology Revolution*. Oxford, Basil Blackwell, 284–94.

Cawkell, A. E. (1980). Forces controlling the paperless revolution. In Forester, T. (ed.), *The Microelectronics Revolution*. Oxford, Basil Blackwell, 244–74.

Child, J. (1972). Organisation structure, environment and performance: the role of strategic choice. *Sociology* **6** (1), 1–22.

(1984). *Organisation: A Guide to Problems and Practice*. 2nd edn. London, Harper and Row.

(1985). Managerial strategies, new technology and the labour process. In Knights, D., Willmott, H. and Collinson, D. (eds), *Job Redesign: Critical Perspectives on the Labour Process*. Aldershot, Gower, 107–11.

Child, J. and Partridge, B. (1982). *Supervisors: The Lost Managers?* Cambridge, Cambridge University Press.

Child, J. *et al*. (1984). Microelectronics and the quality of employment in services. In Marstrand, P. (ed.), *New technology and the Future of Work and Skill*. London, Frances Pinter, 163–90.

Clark, J. (1987). Trade Unions. In Causer, G. (ed.), *Inside British Society*. Brighton, Wheatsheaf, 58–75.

Clark, J. *et al*. (1984). New technology, industrial relations and divisions within the workforce, *Industrial Relations Journal* **15** (3), 36–44.

Clark, J. *et al*. (1988). *The Process of Technological Change: New Technology and Social Choice in the Workplace*. Cambridge, Cambridge University Press.

Clegg, C. and Kemp, N. (1986). Information technology: Personnel where are you? *Personnel Review* **15** (1), 8–15.

Cockburn, C. (1983). *Brothers: Male Dominance and Technological Change*. London, Pluto Press.

(1985a). *Machinery of Dominance: Women, Men and Technical Know-how*. London, Pluto Press.

(1985b). Not my type – choices in technology and organisation for printing. In Collective Design/Projects, *Very Nice Work if You Can Get It: The Socially Useful Production Debate*. Nottingham, Spokesman, 74–83.

(1986). Women and technology: opportunity is not enough. In Purcell, K., Wood, S., Waton, A., and Allen, S. (eds), *The Changing Experience of Employment*. London, Macmillan, 173–87.

Constable, J. and McCormick, R. (1987). *The Making of British Managers*. London, BIM/CBI.

Cooley, M. (1980). The designer in the 1980s: the deskiller deskilled. *Design Studies* **1**, 197–201.

(1981). The social implications of CAD. In Mermet, J. (ed.), *CAD in Medium Sized and*

Small Industries. Amsterdam, North-Holland.

(1985). After the Lucas plan. In Collective Design/Projects, *Very Nice Work if You Can Get It: The Socially Useful Production Debate*. Nottingham, Spokesman, 19–26.

Coombs, R. (1985). Automation, management strategies and labour process change. In Knights, D., Willmott, H. and Collinson, D. (eds), *Job Redesign: Critical Perspectives on the Labour Process*. Aldershot, Gower, 142–70.

Corbett, M. (1985). Prospective work design of a human-centred CNC lathe. *Behaviour and Information Technology* **4** (3), 201–14.

Council for Science and Society, (1981). *New Technology: society, employment, and skill*. London, CSS.

Crompton, R. and Jones, G. (1984). *White Collar Proletariat: Deskilling and Gender in Clerical Work*. London, Macmillan.

CSE Microelectronics Group (1980). *Microelectronics: Capitalist Technology and the Working Class*. London, CSE Books.

Daniel, W. (1987). *Workplace Industrial Relations and Technical Change*. London, Frances Pinter.

Davies, A. (1986). *Industrial Relations and New Technology*. London, Croom Helm.

Dawson, P. M. B. (1986). *Computer Technology and the Redefinition of Supervision*. PhD Thesis, University of Southampton.

(1987a). Computer technology and the job of the first-line supervisor. *New Technology, Work and Employment* **2** (1), 47–60.

(1987b). Information technology and the control function of supervision. In Knights, D. and Willmott, H. (eds), *New Technology and the Labour Process*. London, Macmillan, 152–74.

Dawson, P. M. B. and McLoughlin, I. P. (1986). Computer technology and the redefinition of supervision. *Journal of Management Studies* **23** (1), 116–32.

(1988) Organisational choice in the redesign of supervisory systems. In Boddy, D. McCalman, J. and Buchanan, D. (eds.), *The New Management Challenge*. London, Croom Helm 99–110.

Dawson, S. (1986). *Analysing Organisations*. London, Macmillan.

Dawson, S. and Wedderburn, D. (1980). Introduction. In Woodward, J. *Industrial Organisation: Theory and Practice*, 2nd cdn. Oxford, Oxford University Press.

Dickinson, M. (1984) *To Break a Union: the Messenger, the State and the NGA*. Manchester, Booklist.

Dodgson, M. and Martin, R. (1987). Trade union policies on new technology: facing the challenge of the 1980s. *New Technology, Work and Employment* **2** (1), 9–18.

Downing, H. (1980). Word processors and the oppression of women. In Forester, T. (ed.), *The Microelectronics Revolution*. Oxford, Basil Blackwell, 275–87.

(1982). On being automated. *Aslib Proceedings* **35** (1), January, 38–51.

Dunn, S. (1985). The law and the decline of the closed shop in the 1980s. In Fosh, P. and Littler, C.R., *Industrial Relations and the Law in the 1980s: Issues and Future Trends*. Aldershot, Gower, 82–117.

Dunn, S. and Gennard, J. (1984). *The Closed Shop in British Industry*. London, Macmillan.

Edwards, R. (1979). *Contested Terrain: The Transformation of Work in the Twentieth Century*. London, Heinemann.

Eldridge, J. E. T. (1983). Review of Wood, S. (ed.), *The Degradation of Work?*, and Littler, C.R., *The Development of the Labour Process in Capitalist Societies*. In *British Journal of Industrial Relations* **25** (1), 418–20.

Elliott, D. (1986). The GLC's innovation and employment initiative. Open University Technology Policy Group Occasional Paper, No.11.

Elliott, J. (1978). *Conflict or Co-operation? The Growth of Industrial Democracy*. London, Kogan Page.

Euromonitor (1985). *Electronic Developments in Retailing*. Retail and Distribution Surveys. London, Euromonitor Publications.

Feigenbaum, E. and McCorduck, P. (1985). Land of the rising fifth generation. In Forester, T. (ed.), *The Information Technology Revolution*. Oxford, Basil Blackwell, 71–83.

Forester, T. (1987). *High-Tech Society*. Oxford, Basil Blackwell.

Forester, T. (ed.) (1980). *The Microelectronics Revolution*. Oxford, Basil Blackwell.

(1985). *The Information Technology Revolution*. Oxford, Basil Blackwell.

Fox, A. (1966). *Industrial Sociology and Industrial Relations*. Research Paper no. 3, Royal Commission on Trade Unions and Employers Associations. London, HMSO.

(1985). *Man Mismanagement*. 2nd edn. London, Hutchinson.

Francis, A. (1986). *New Technology at Work*. Oxford, Oxford University Press.

Francis, A. *et al.* (1982). The impact of information technology at work: the case of CAD/CAM and MIS in engineering plants. In Bannon, L., Barry, U. and Holst, O. (eds), *Information Technology: Impact on a Way of Life*. Dublin, Tycooly, 182–96.

Freeman, C. (1986). The diffusion of innovations – microelectronics technology. In Roy, R. and Wield, D. (eds.), *Product Design and Technological Innovation*. Milton Keynes, Open University Press, 193–200.

Freeman, C. *et al.* (1982). *Unemployment and Technical Innovation*. London, Frances Pinter.

Friedman, A. (1977). *Industry and Labour*. London, Macmillan.

Friedrichs, G. and Schaff, A. (1982). *Microelectronics and Society: For better or for worse?* Report to Club of Rome. Elmsford, Pergamon.

Gallie, D. (1978). *In Search of the New Working Class*. Cambridge, Cambridge University Press.

Gennard, J. (1987). The NEA and the impact of new technology. New Technology, Work and Employment **2** (2), 126–41.

Gennard, J. and Dunn, S. (1983). The impact of new technology on the structure and organisation of craft unions in the printing industry. *Brit. J. Industrial Relations* **XXI** (1), 17–32.

Gill, C. (1985). *Work, Unemployment and New Technology*. Cambridge, Polity Press.

Giuliano, V. E. (1985). The mechanisation of office work. In Forester, T. (ed.), *The Information Technology Revolution*. Oxford. Basil Blackwell, 298–311.

Gourlay, S. (1987). The design of work organisation: or what do work organisers do? Paper presented at British Sociological Association Annual Conference, University of Leeds, April.

Greve, R. M. (1986). The effects of technological change on women workers. *Technological Change and Labour Relations*. Proc. 7th World Congress Int. Indust. Rel. Assoc., Hamburg, Vol.1, 171–82.

Hage, J. (1980). *Theories of Organisation*. New York, John Wiley.

Hall, G. (1986). The geography of the 5th Kondratieff cycle. In Roy, R. and Wield, D. (eds.), *Product Design and Technological Innovation*. Milton Keynes, Open University Press, 265–70.

Handy, C. (1987). *Managers in Five Countries: a new professionalism*. A report for the NEDC, MSC and BIM, London, National Economic Development Office.

Hartmann, G. *et al.* (1985). Computerised machine tools, manpower consequences and skill utilisation. In Rhodes, E. and Wield, D. (eds), *Implementing New Technologies: Choice, Decision, and Change in Manufacturing*. Oxford, Basil Blackwell, 352–60.

Hill, S. (1981). *Competition and Control at Work*. London, Hutchinson.

Hunter, L. C. *et al.* (1970). *Labour Problems of Technological Change*. Edinburgh, George Allen and Unwin.

Huws, U. (1982). *Your Job in the Eighties: a Woman's Guide to New Technology*. London, Pluto Press.

(1984). *The New Homeworkers: New Technology and the Changing Location of Office Work*. Low Pay Unit, Pamphlet No. 28., London, Low Pay Unit.

Institute of Manpower Studies (1984). Flexible manning – the way ahead. IMS Report No. 88, Sussex University, IMS and Manpower Ltd.

Jacobs, A. (1983). *Film and Electronic Technologies in the Production of Television News*. PhD Thesis, University of Southampton.

Jary, S. (1987) Negotiating technological change?: bargaining power, union resoures and workplace organisation. Paper presented at British Sociological Association Annual Conference, University of Leeds, April.

Jonas, W. (1988). For a more comprehensive view of CAD. In Rader, M., Wingert, B. and Riehm, U. (eds.), *Social Science Studies of CAD/CAM*. Heidelberg, Physica-Verlag, 275–85.

Jones, B. (1982). Distribution or redistribution of engineering skills? The case of numerical control. In Wood, S. (ed.), *The Degradation of Work?: Skill, Deskilling and the Labour Process*. London, Hutchinson, 179–200.

(1985). Technical, organisational and political constraints on system re-design for machinist programming of NC machine tools. Paper presented at IFIP Conference on 'System Design for the Users', Italy, September.

Jones, B. and Scott, P. (1987). Flexible manufacturing systems in Britain and the USA. *New Technology, Work and Employment* **2** (1), 27–36.

Kaplinsky, R. (1982). *Computer Aided Design: Electronics, Comparative Advantage and Development*. London, Frances Pinter.

(1984). *Automation: the Technology and Society*. London, Longman.

Kinnie, N. and Arthurs, A. (1986). New techniques for recording time at work: their implications for supervisory training and development. In Boddy, D., McCalman, J. and Buchanan, D. (eds.), *The New Management Challenge*. London, Croom Helm, 111–23.

Knights, D. and Willmott, H. (eds). (1986a). *Managing the Labour Process*. Aldershot, Gower.

(1986b). *Gender and the Labour Process*. Aldershot, Gower.

(1987). *New Technology and the Labour Process*. London, Macmillan.

Knights, D., Willmott, H. and Collinson, D. (eds). (1985) *Job Redesign: Critical Perspectives on the Labour Process*. Aldershot, Gower.

Leach. B. and Shutt, J. (1985). Chips and crisps: labour facing the crunch. In Forester, T. (ed.), *The Information Technology Revolution*. Oxford, Basil Blackwell, 480–95.

Leavitt, H. J. and Whisler, T. L. (1958). Management in the '80s. *Harvard Business Review* **36** (6), 41–8.

Lee, D. (1982). Beyond deskilling: skill, craft and class. In Wood, S. (ed.), *The Degradation of Work?: Skill, Deskilling and the Labour Process*. London, Hutchinson, 146–62.

Levie, H. and Williams, R. (1983). User involvement and industrial democracy: problems and strategies in Britain. In Briefs, U., Ciborra, C. and Schneider, L. (eds), *Systems Design For, With and by the Users*. Amsterdam North-Holland, 265–404.

Littler, C. R. (1982). *The Development of the Labour Process in Capitalist Societies*. London, Heinemann.

(1985). Taylorism, Fordism and job design. In Knights, D., Willmott, H., and Collinson, D. (eds), *Job Redesign: Critical Perspectives on the Labour Process*. Aldershot, Gower, 10–29.

Littler, C. R. and Salaman, G. (1984). *Class at Work*. London, Batsford.

Lyon, D. (1986). From 'Post-industrialism' to 'Information Society': a new social transformation? *Sociology* **20** (4), 577–88.

Mackenzie, D. and Wajcman, J. (1985). Introductory Essay. In MacKenzie, D. and Wajcman, J. (eds) (1985), *The Social Shaping of Technology*. Milton Keynes, Open University Press, 2–25.

Mackenzie, D. and Wajcman, J. (eds) (1985). *The Social Shaping of Technology*. Milton Keynes, Open University Press.

Macrae, N. (1982). 'Intrapreneurial Now'. *The Economist,* 17 April, 47–52.

Manwaring, T. (1981). The trade union response to new technology. *Industrial Relations Journal* **12** (4), 6–27.

Marginson, P. *et al.* (eds.) (1988). *Beyond the Workplace.* Oxford, Basil Blackwell.

Markey, R. (1982). New technology, the economy and the unions in Britain. *Journal of Industrial Relations,* December.

Marstrand, P. (ed.) (1984). *New Technology and the Future of Work and Skill.* London, Frances Pinter.

Marti, J. and Zeilinger, A. (1985). New technology in banking and shopping. In Forester, T. (ed.), *The Information Technology Revolution.* Oxford, Basil Blackwell, 350–8.

Martin, J. (1978). *The Wired Society.* Harmondsworth, Penguin.

Martin, R. (1981). *New Technology and Industrial Relations in Fleet Street.* Oxford, Oxford University Press.

—— (1984). New technology and industrial relations in Fleet Street. In Warner, M. (ed.) (1984). *Microprocessors, Manpower and Society.* Aldershot, Gower, 240–52.

—— (1988). The management of industrial relations and new technology. In Marginson, P. *et al.* (eds.), *Beyond the Workplace.* Oxford, Basil Blackwell.

Mayo, J. S. (1985). The evolution of the intelligent network. In Forester, T. (ed.), *The Information Technology Revolution.* Oxford, Basil Blackwell, 106–19.

McLoughlin, I. P. (1983a). *Industrial Engineers and Theories of the New Middle Class.* PhD Thesis, University of Bath.

—— (1983b). Problems of management control and the introduction of new technology. Working Paper, New Technology Research Group, University of Southampton.

—— (1986). *Innovation by Design.* Research Report, University of Southampton.

—— (1987). The Taylorisation of intellectual work?: the case of CAD. Paper presented at British Sociological Association Annual Conference, University of Leeds, April.

—— (1988). Management Strategies for the introduction and control of interactive computer graphics systems. In Rader, M., Wingert, B. and Riehm, U. (eds.), *Social Science Studies of CAD/CAM in Europe.* Heidelberg, Physica-Verlag, 133–43.

McLoughlin, I. P., Rose, H. and Clark, J. (1985). Managing the introduction of new technology. *Omega* **13** (4), 251–62.

McLoughlin, I. P., Smith, J. H. and Dawson, P. M. B. (1983). *The Introduction of a Computerised Freight Information System in British Rail-TOPS.* Research Report, New Technology Research Group, University of Southampton.

Melvern, L. (1986). *The End of the Street.* London, Methuen.

Mintzberg, H. (1978). Patterns in strategy formation. *Management Science* **14** (9), 934–48.

Moore, R. and Levie, H. (1985). New technology and the unions. In Forester, T. (ed.), *The Information Technology Revolution.* Oxford, Basil Blackwell.

Mortimer, J. E. (1971). *Trade Unions and Technological Change.* Oxford, Oxford University Press.

Murray, F. (1987). Women, trade union technology bargaining, and the need for a new strategy. Human Centred Office System Project Working Paper, Sheffield Polytechnic.

National Computing Centre, (1986). *The Impact of Office Technology.* Manchester, NCC.

Noble, D. (1979). Social choice in machine design: the case of automatically controlled machine tools. In Zimbalist, A. (ed.), *Case Studies in the Labour Process.* New York, Monthly Review Press, 18–50.

—— (1984). *Forces of Production: A Social History of Industrial Automation.* New York, Alfred A. Knopf.

Northcott, J. (1986). *Microelectronics in Industry: promise and performance.* London, Policy Studies Institute.

Northcott, J. and Rogers, P. (1984). *Microelectronics in British Industry: Patterns of Change.* London, Policy Studies Institute.

(1985). *Chips and Jobs.* London, Policy Studies Institute.

Northcott, J. and Rogers, P. with Zeilinger, A. (1982). *Microelectronics in Industry: Survey Statistics.* London, Policy Studies Institute.

Noyce, R. (1980). Microelectronics. In Forester, T. (ed.), *The Microelectronics Revolution.* Oxford, Basil Blackwell, 29–61.

Open University T362 Course Team (1986a). *Innovation Waves,* T362, Block 5, Unit 13. Milton Keynes, Open University Press.

(1986b). *Government,* T362, Block 4, Units 11–12. Milton Keynes, Open University Press.

Penn, R. (1982). Skilled manual workers in the labour process, 1856–1964. In Wood, S. (ed.), *The Degradation of Work?: Skill, Deskilling and the Labour Process.* London, Hutchinson, 90–121.

Pollert, A. (1987). *The Flexible Firm: a Model in Search of Reality or a Policy in Search of a Practice?* Warwich Papers in Industrial Relations.

Potter, S. (1987). *On the Right Lines? The Limits to Technological Innovation.* London, Frances Pinter.

Preece, D. A. (1986). Organisations, flexibility and new technology. In Voss, C. A. (ed.), *Managing Advanced Manufacturing Technology.* Berlin, Springer-Verlag.

Preece, D. A. and Harrison, M. R. (1986). The contribution of personnel specialists to technology related organisational change. Unpublished paper.

Pugh, D. S. and Hickson, D. J. (1976). *Organisation Structure in its Context: The Aston Programme I.* Farnborough, Saxon House.

Purcell, J. and Sisson, K. (1983). Strategies and practice in the management of industrial relations. In Bain, G. (ed.), *Industrial Relations in Britain.* Oxford, Basil Blackwell, 95–120.

Rada, J. (1980). *The Impact of Microelectronics.* Geneva, International Labour Office.

Rajan, A. (1984). *New Technology and Employment in the Financial Services Sector.* Aldershot, Gower.

Ray, G. (1986). Innovation in the Long Cycle. In Roy, R. and Wield, D. (eds.), *Product Design and Technological Innovation.* Milton Keynes, Open University Press, 271–8.

Reeves, T. K., Turner, B. A. and Woodward, J. (1970). Technology and organisational behaviour. In Woodward, J. (ed.), *Industrial Organisation: Behaviour and Control.* Oxford, Oxford University Press, 3–18.

Reeves, T. K. and Woodward J. (1970). The study of managerial control. In Woodward, J. (ed.) (1970), *Industrial Organisation: Behaviour and Control.* Oxford, Oxford University Press, 37–56.

Rhodes, E. and Wield, D. (eds) (1985). *Implementing New Technologies: Choice, Decision, and Change in Manufacturing.* Oxford, Basil Blackwell.

Rice, A. (1958). *Productivity and Social Organisation.* London, Tavistock.

Robbins, K. and Webster, F. (1982). New technology: a survey of trade union response in Britain. *Industrial Relations Journal* **13** (2), 7–26.

Robey, D. (1977). Computers and management structure. *Human Relations* **30** (9), 963–76.

Rolfe, H. (1986). Skill, de-skilling, and new technology in the non-manual labour process. *New Technology, Work and Employment* **1** (1), 18–26.

(1987). *New Technology, Skill and Deskilling in Non-Manual Work.* PhD Thesis, University of Southampton.

Rose, H. *et al.* (1986). Opening the black box: the relation between technology and work. *New Technology, Work and Employment* **1** (1), 18–26.

Rose, M. (1978). *Industrial Behaviour: Theoretical Development since Taylor.* London, Penguin.

Rosenbrock, H. H. (1985a). Engineers and the work that people do. In Littler, C.R. (ed.), *The Experience of Work.* Aldershot, Gower, 161–71.

(1985b). Engineering design and social science. Discussion paper for ESRC/SPRU Workshop on New Technology in Manufacturing Industry. Windsor, May.

Rothwell, S. (1984). Company employment policies and new technology in manufacturing and service sectors. In Warner, M. (ed.), *Microelectronics, Manpower and Society*. Aldershot, Gower, 111–33.

(1985). Supervisors and new technology. In Rhodes, E. and Wield, D. (eds), *Implementing New Technologies: Choice, Decision, and Change in Manufacturing*. Oxford, Basil Blackwell, 374–83.

Routledge, P. (1979). The dispute at Times Newspapers Ltd.: a view from inside. *Industrial Relations Journal* **10** (4), 5–9.

Roy, R. and Wield, D. (eds.) (1986). *Product Design and Technological Innovation*. Milton Keynes, Open University Press.

Rush, H. and Williams, R. (1984). Consultation and change: new technology and manpower in the electronics industry. In M. Warner (ed.), *Microelectronics, Manpower and Society*. Aldershot, Gower, 171–88.

Senker, P. (1985a). Some lessons from research. In Senker, P. (ed.), *Planning for Microelectronics in the Workplace*. Aldershot, Gower 155–71.

(1985b). Implications of CAD/CAM for management. In Rhodes, E. and Wield, D. (eds) *Implementing New Technologies*. Oxford, Basil Blackwell, 225–34.

Silverman, D. (1970). *The Theory of Organisations*. London, Heinemann.

Sleigh, J. *et al.* (1979). *The Manpower Implications of Microelectronics*. Department of Employment London, HMSO.

Smith, S. L. (1987a). Information Technology: Taylorisation or human-centred office systems. *Science and Public Policy* **14** (3). 159–67.

(1987b). How much change in store? The impact of new technologies on managers and staffs in retail distribution. In Knights, D. and Willmott, H. (eds), *New Technology and the Labour Process*. London, Macmillan.

Smith, S. L. and Wield, D. (1987). New technology and bank work: banking on I.T. as an 'organisational technology'. In Harris, L. (ed.), *New Perspectives on the Financial System*. London, Croom Helm.

Somers, G. G., Cushman, L. and Wienberg, N. (eds) (1963). *Adjusting to Technological Change*. Westport, Conn., Greenwood Press.

Sorge, A. *et al.* (1983). *Microelectronics and Manpower in Manufacturing Applications of Computer Numerical Control in Great Britian and West Germany*. Aldershot, Gower.

Steffens, J. (1983). *The Electronic Office: Progress and Problems*. London, Policy Studies Institute.

Steiber, J. (ed.) (1966). *Employment Problems of Automation and Advanced Technology*. London, Macmillan.

Stonier, T. (1983). *The Wealth from Information*. London, Thames and Hudson.

Storey, J. (1983). *Managerial Prerogative and the Question of Control*. London, Routledge and Kegan Paul.

(1986). The Phoney War? New office technology: organisation and control. In Knights, D. and Willmott, H. (eds), *Managing the Labour Process*. Aldershot, Gower, 44–66.

Strauss, A. *et al.* (1973). The hospital and its negotiated order. In Salaman, G. and Thompson, K. (eds), *People and Organisations*. London, Longman, 303–320.

Swann, J. (1986). *The Employment Effects of Microelectronics in the UK Service Sector*. London, Technical Change Centre.

Taber-Heyder, H. (1985). The impact of CAD systems on professional practices. Unpublished paper.

Thomason, G. (1981). *A Textbook of Personnel Management*. Fourth edn., London, Institute of Personnel Management.

Thompson, P. (1983). *The Nature of Work*. London, Macmillan.

Thompson, P. and Bannon, E. (1985). *Working the System: The Shop Floor and New Technology*. London, Pluto Press.

Thurley, K. (1966). Changing technology and the supervisor. In Steiber, J. (ed.), *Employment Problems of Automation and Advanced Technology.* London, Macmillan.

Thurley, K. and Wirdenius H. (1973). *Supervision: A Reappraisal.* London, Heinemann.

Toffler, A. (1980). *The Third Wave.* London, Pan Books.

Trades Union Congress (1974). *Industrial Democracy.* London, TUC.

(1979). *Employment and Technology.* London, TUC.

(1984). *Women and New Technology,* a TUC discussion document. London, TUC.

Turner, H. A. (1962). *Trade Union Growth, Structure and Policy.* London, Allen and Unwin.

Wainwright, H. and Elliott, D. A. (1982). *The Lucas Plan: A New Trade Unionism in the Making?* London, Allison and Busby.

Walton, R. (1986). Social choice in the development of advanced information technology. In Rhodes. E. and Wield, D. (eds), *Implementing New Technologies: Choice, Decision, and Change in Manufacturing.* Oxford, Basil Blackwell, 345–51.

Warner, M. (ed.) (1984). *Microprocessors, Manpower and Society.* Aldershot, Gower.

Watson, T. J. (1977). *The Personnel Managers.* London, Routledge and Kegan Paul.

(1986). *Management Organisation and Employment Strategy.* London, Routledge and Kegan Paul.

Webster, J. (1986). Word processing and the secretarial labour process. In Purcell, K., Wood, S., Waton, A. and Allen, S. (eds), *The Changing Experience of Employment.* London, Macmillan, 114–31.

Weizenbaum, J. (1985). The myths of artificial intelligence. In Forester, T. (ed.), *The Information Technology Revolution.* Oxford, Basil Blackwell, 84–94.

Werneke, D. (1985). Women; the vulnerable group. In Forester, T. (ed.), *The Information Technology Revolution.* Oxford, Basil Blackwell, 400–16.

West, J. (1982). New technology in women's office work. In West, J. (ed.). *Work, Women and the Labour Market.* London, Routledge and Kegan Paul, 61–79.

Wilkinson, B. (1983). *The Shop Floor Politics of New Technology.* London, Heinemann.

(1985). The politics of technical change. In Forester, T. (ed.), *The Information Technology Revolution.* Oxford, Basil Blackwell, 439–53.

Williams, R. and Steward, F. (1985). New technology agreements: an assessment. *Industrial Relations Journal* **16** (3), 58–73.

Willman, P. (1986). *Technological Change, Collective Bargaining and Industrial Efficiency.* Oxford, Oxford University Press.

Willman, P. (1987). *New Technology and Industral Relations: a Review of the Literature.* Department of Employment Research Paper No. 56. London, DOE.

Willman, P. and Cowan, R. (1984). New technology in banking: the impact of autotellers on staff numbers. In Warner, M. (ed.) (1984). *Microprocessors, Manpower and Society.* Aldershot, Gower, 189–239.

Willman, P. and Winch, G. (1985). *Innovation and Management Control.* Cambridge, Cambridge University Press.

Winch, G. (ed.) (1983). *Information Technology in Manufacturing Processes.* London, Rossendale.

Winner, L. (1977). *Autonomous Technology.* Cambridge, Mass., MIT Press.

(1985). Do artifacts have politics? In Mackenzie, D. and Wajcman, J. (eds) (1985). *The Social Shaping of Technology.* Milton Keynes, Open University Press, 26–38.

Winstanley, D. and Francis, A. (1987). Redrawing the line – changing design practices in engineering firms. Paper presented at British Sociological Association Annual Conference, University of Leeds, April.

Winterton, J. and Winterton, R. (1985). *New Technology: The Bargaining Issues.* Leeds–Nottingham Occasional Paper No. 7. Universities of Leeds and Nottingham/Institute of Personnel Management.

Wood, S. (ed.) (1982). *The Degradation of Work?: Skill, Deskilling and the Labour Process.* London, Hutchinson.

Wood, S. and Kelly, J. (1982). Taylorism, responsible autonomy and management strategy. In Wood, S. (ed.) (1982). *The Degradation of Work?: Skill, Deskilling and the Labour Process.* London, Hutchinson, 74–89.

Woodward, J. (ed.) (1970) *Industrial Organisation: Behaviour and Control.* Oxford, Oxford University Press.

(1980). *Industrial Organisation: Theory and Practice,* 2nd ed. Oxford, Oxford University Press.

Zimbalist, A. (ed.) (1979). *Case Studies on the Labour Process.* New York, Monthly Review Press.

Index